Distillation tray fundamentals

Trial assembly of large two-pass slotted sieve trays. (Courtesy of Union Carbide Corporation.)

Distillation tray fundamentals

M.J.LOCKETT
Union Carbide Corporation

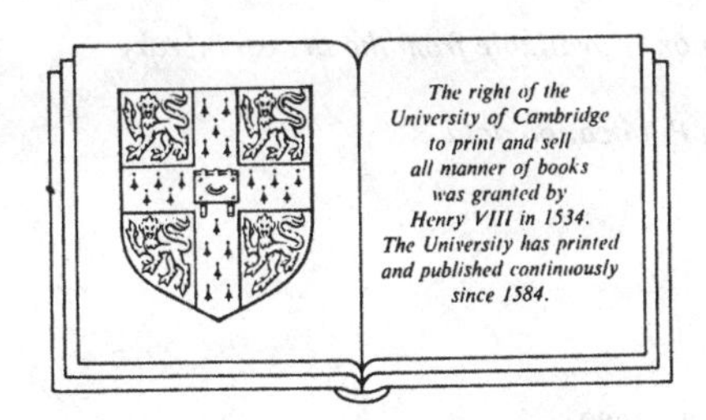

CAMBRIDGE UNIVERSITY PRESS
Cambridge
London New York New Rochelle
Melbourne Sydney

CAMBRIDGE UNIVERSITY PRESS
Cambridge, New York, Melbourne, Madrid, Cape Town, Singapore, São Paulo, Delhi

Cambridge University Press
The Edinburgh Building, Cambridge CB2 8RU, UK

Published in the United States of America by Cambridge University Press, New York

www.cambridge.org
Information on this title: www.cambridge.org/9780521105873

First published 1986
This digitally printed version 2009

A catalogue record for this publication is available from the British Library

Library of Congress Cataloguing in Publication data

Lockett, M. J.
Distillation tray fundamentals.

Bibliography
Includes index.
1. Distillation apparatus. I. Title.
TP159.D5L63 1986 660.2′8425 85-30891

ISBN 978-0-521-32106-8 hardback
ISBN 978-0-521-10587-3 paperback

Contents

Preface

The design and performance of distillation trays is one of the most thoroughly studied topics in chemical engineering. Papers dealing with trays are available in abundance but the information they contain is often conflicting. Indeed, by searching hard enough it is usually possible to find some published information to support nearly any contention one cares to make about how trays work. One aim of this book has been to draw together and interpret much of this previous work.

It has not been my intention to write a design manual for tray design. Others have done this far more effectively than I, and some such sources are cited in Chapter 1. The approach I have taken is to assume that the reader is already familiar with elementary design methods as given in the standard undergraduate textbooks. What I have attempted is to delve a little deeper into the empirical correlations and equations which are used for tray design and to indicate their origins and shortcomings. The sequence of the material covered is such that, when coupled with the outline design procedure given in Chapter 1, it is relatively straightforward to adapt it to and improve existing design methods. Having said all that, I suppose this book was mainly written for the same reason that most books of its type are written – having worked in an area, in this case distillation trays, for some time with considerable enjoyment, it felt appropriate to convey my enthusiasm for the subject to others.

Distillation trays are seemingly so simple that it may be surprising that they warrant a whole book, albeit a small one, to themselves. Sieve tray decks are, after all, hardly more than sheets of metal with a few holes punched in them. This of course is part of the fascination – that the behaviour of something so simple can be so difficult to predict with regard to its hydrodynamic and mass transfer performance. Another equally intriguing aspect is the interplay between the topics covered in this book and the marketing and selling of trays. Distillation column internals are marketed very competitively as a glance at the advertisements in the popular chemical engineering magazines will testify. It is necessary to be aware that however erudite and interesting in their own right are theories of tray behaviour, they must in the end be translated into the design of pieces of metal which work reliably and at reasonable cost. Thus, those involved with trays span a wide range from those concerned solely with metal fabrication through to researchers into fairly esoteric irreversible thermodynamics. This complex mix of concerns and priorities also provides much of the subject's interest.

It is hoped that most people who deal with distillation trays at any level will find something of value in the pages that follow. However, the book is primarily aimed at the process engineer involved with design who wishes to know more than can be obtained by simply applying correlations. In addition, those concerned with specifying and buying trays should find it useful, as should those troubleshooting malfunctioning columns. I know from experience that it forms a reasonable basis for a graduate course, and the lack of answers to many of the questions posed should also provide a reasonable stimulus for research workers in this area of chemical engineering.

The book had its origins in a course of lectures I gave on distillation in the late 1970s to graduate students at the University of Manchester Institute of Science and Technology. When I was asked to give a course on distillation trays at the University of Aston to practising engineers as part of the continuing education series of the Institution of Chemical Engineers, the idea took hold of developing the material into a book. A further impetus was provided when I joined the Linde Division of Union Carbide, responsible for distillation engineering and development. This helped me put some of the more theoretical material into context as it impacts on the practising engineer.

Although the book was started while I was at UMIST, 95% of it was completed at nights and weekends while working for Union Carbide. In retrospect I doubt whether I would have started it if I had known it would be written outside the cloistered life of the university. In the event, my wife and children were extremely patient and understanding during its completion in spite of my absence from numerous school concerts, baseball games and skiing trips.

There are of course many other people to whom acknowledgement is due. In particular I would like to thank K E Porter who introduced me to the subject and taught me the importance of asking the right questions about research problems. I also appreciated working with the late G L

Standart, who impressed on me the other aspect – the importance of giving the right answers! A source of inspiration has been the research students with whom I have had the good fortune to work over the years, and this book is built substantially on their efforts. Colleagues who have read the manuscript and contributed suggestions include M R Resetarits and D R Summers to whom thanks are also due, although errors and omissions are of course solely my own responsibility. Mrs Anita Strzalkowski did a sterling job of transforming my ill-written scrawl into a typewritten manuscript with good-humour and patience, and J D Augustyniak drew the figures with his usual skill and attention to detail. Finally I would like to thank the directors of Union Carbide for permission to publish the book.

Acknowledgements

The author is grateful to the following organisations for permission to reproduce or adapt figures which have been published by other authors. American Institute of Chemical Engineers: Figs. 4.3 and 9.20; Energy Publications: Figs. 5.1 and 5.6; Glitsch, Inc.: Fig. 5.2; Institution of Chemical Engineers: Figs. 1.3, 1.6, 2.2, 2.4, 2.8, 2.12, 2.18, 2.20, 2.21, 2.22; Pergamon Press, Ltd.: Figs. 2.3, 2.7, 2.11, 2.13, 2.14; VCH Verlagsgesellschaft: Figs. 4.2 and 5.7.

Nomenclature

Unless otherwise stated locally, the units used are as given below. In particular, the following units should be used in equations which are not dimensionless. The nomenclature of the original sources has been retained as far as possible. The terms gas and vapour are used interchangeably.

A	Cross-sectional area, m^2
A	Constant in eqn. (4.11)
A	Defined by eqn. (9.32)
A_b	Tray bubbling area, m^2
A_d	Downcomer cross-sectional area, m^2
A'_{da}	Area under splash baffle, m^2
A_h	Area of holes in tray deck, m^2
A_n	Net area for liquid disengagement above tray, m^2
A_{sc}	Curtain area of 'closed' valves, m^2
A_V	Area of valve disk, m^2
a	Interfacial area per unit volume of two-phase dispersion, $m^2\,m^{-3}$
a	Parameter describing jet shape, eqn. (2.17)
a	Parameter in eqn. (2.31)
a	Defined by eqns. (10.8)
a'	Interfacial area per unit volume of vapour, $m^2\,m^{-3}$
$\bar{a}$	Interfacial area per unit volume of liquid, $m^2\,m^{-3}$
B	Defined by eqns. (9.32)
B_s	Oscillation number, eqn. (5.23)
b	Intercept of equilibrium line (binary)
b	Constant in eqn. (4.16)
b	Defined by eqns. (10.8)
C	Defined by eqn. (3.25)
C	Defined by eqn. (9.32)
C_d	Discharge coefficient

C_0	Drop drag coefficient
CF	Capacity factor based on A_b, m s^{-1}
CF'	Capacity factor based on A_n, m s^{-1}
CF''	Capacity factor from Fig. 5.1, m s^{-1}
CF_{max}	Theoretical maximum capacity factor, eqn. (5.10), m s^{-1}
CF_0	Capacity factor at zero liquid load, m s^{-1}
c	Velocity of sound, m s^{-1}
D	Column diameter, m
D	Defined by eqn. (9.32)
D_G, D_L	Diffusion coefficient in vapour and liquid phase, m^2 s^{-1}
D_L^0	Diffusion coefficient in liquid phase at infinite dilution, m^2 s^{-1}
De	Eddy diffusivity for liquid mixing, m^2 s^{-1}
De_G	Eddy diffusivity for vapour mixing, m^2 s^{-1}
d_b	Bubble diameter, m
d_{bs}	Sauter mean bubble diameter, m
d_h	Hole diameter, m
d_j	Jet diameter, m
d_p	Drop diameter, m
d_{pm}	Mean drop diameter, m
d_{ps}	Sauter mean projected drop diameter, m
$d_p(j)$	Diameter of drops in subgroup j, m
E^+	Liquid entrainment rate, kg-mol s^{-1}
E_H	Hausen tray efficiency
E_{HL}, E_{HV}	Hausen liquid and vapour tray efficiency
E_m	Liquid entrainment rate, kg s^{-1}
E_{ML}	Murphree liquid phase tray efficiency
E_{MV}	Murphree vapour phase tray efficiency
E_{MV}^a	Apparent Murphree vapour phase tray efficiency
E_{ni}^v	Vaporisation tray efficiency for component i on tray n
E_0	Section efficiency
E_{OG}	Murphree vapour phase point efficiency
E_S	Standart material efficiency
E_{SH}	Standart enthalpy efficiency
E_{Si}	Standart component efficiency
E_{TL}, E_{TV}	Thermal efficiency in liquid and vapour phase
e	Liquid entrainment rate ÷ vapour rate, (kg-mol s^{-1}) ÷ (kg-mol s^{-1})
e_m	Liquid entrainment rate ÷ vapour rate, (kg s^{-1}) ÷ (kg s^{-1})
e_0	E^+/L_0
F	Defined by eqn. (9.39)
$F(j)$	Number fraction of drops having size $d_p(j)$
F_h	Hole F factor, $u_h\rho_G^{0.5}$, kg$^{0.5}$ m$^{-0.5}$ s^{-1}
$(F_h)_{crit}$	Critical value of hole F factor

F_R Reaction at the plate, N
F_s Superficial F factor, $u_s \rho_G^{0.5}$, $kg^{0.5}\ m^{-0.5}\ s^{-1}$
F_w Frictional force due to walls, N
F_x Force, N
FF Flood factor or fractional approach to flooding
FP Flow parameter, eqn. (1.1)
Fr Froude number, u_s^2/gh_{cl}
Fr' Modified Froude number, $Fr\ \rho_G/(\rho_L - \rho_G)$
Fr_1 Froude number for liquid flow, $u_{L1}/(gh_1)^{0.5}$
Fr_h Froude number, $u_h[\rho_G/gh_{cl}(\rho_L - \rho_G)]^{0.5}$
f Friction factor for froth flow
f_i Fraction of vapour flow carried by bubbles of species i
$\bar{f}_{ni}^L, \bar{f}_{ni}^V$ Fugacity of component i in liquid and vapour leaving tray n, bar

G Vapour flow rate, kg-mol s^{-1}
G Defined by eqn. (9.39)
G' Vapour flow rate per unit bubbling area, kg-mol $s^{-1}\ m^{-2}$
G_{ij} Matrix coefficients defined by eqn. (10.8)
g Acceleration due to gravity, m s^{-2}

H Vapour enthalpy, kJ $(\text{kg-mol})^{-1}$
H Defined by eqn. (9.39)
h Liquid enthalpy, kJ $(\text{kg-mol})^{-1}$
h Distance from tray floor, m
$\Delta h(i)$ Height increment at height $h(i)$, m
h_{cl} Clear liquid height, m
h_{clD} Dynamic liquid head at tray floor, m
$(h_{cl})_{ow}$ Height of clear liquid flowing over weir, m
h_{cli} Clear liquid height at liquid entry, m
h_{co} Height of clear liquid in exit calming zone, m
h'_{da} Head loss for liquid flowing under splash baffle, m of liquid
h_{DT} Dry tray pressure drop, m of liquid
h'_{DT} $h_{DT} + h_R$, m of liquid
h_f Height of two-phase dispersion on tray, m
h_{fd} Froth height in downcomer, m
h_G, h_L Heat transfer coefficients in vapour and liquid phases, kW $m^{-2}\ K^{-1}$
h_i Depth of liquid at liquid entry, m
h_j Jet height, m
h_L Depth of liquid in vessel, m
h_m Head of liquid measured by manometer, m of liquid
h_n Pressure increase across the nappe, m of liquid
h_{ow} Height of froth flowing over weir, m
h_R Residual pressure drop, m of liquid
h_{udc} Pressure drop for flow under downcomer, m of liquid
h_w Weir height, m

h_{WT} Wet (or total) tray pressure drop, m of liquid
h_1 Clearance under downcomer, m
h_1, h_2 Depth of liquid at positions 1 and 2, m

i Height increment number

j Drop subgroup number

K Coefficient in eqn. (4.20)
K_{ni} Ideal solution K value for component i on tray n
K_{OG} Overall mass transfer coefficient based on vapour, kg-mol s^{-1} m^{-2}
K'_{OG} Overall mass transfer coefficient based on vapour, m s^{-1}
K_w Parameter used in eqn. (4.10)
K_1, K_2 Constants in eqn. (5.17)
k Correction factor, eqn. (3.17)
k' Correction factor, eqn. (3.32)
k_G Vapour phase mass transfer coefficient, kg-mol s^{-1} m^{-2}
k'_G Vapour phase mass transfer coefficient, m s^{-1}
k_L Liquid phase mass transfer coefficient, kg-mol s^{-1} m^{-2}
k'_L Liquid phase mass transfer coefficient, m s^{-1}

L Liquid flow rate, kg-mol s^{-1}
L_0 Liquid flow rate on tray in absence of entrainment or weeping, Figs. 9.11, 9.14, kg-mol s^{-1}
L_w Weeping rate, kg-mol s^{-1}
L'_w Weeping rate per unit tray area, kg-mol s^{-1} m^{-2}
l Latent heat, kJ $(\text{kg-mol})^{-1}$
l Maximum valve lift, mm
l Parameter in eqn. (2.32), m

M Molecular weight, kg $(\text{kg-mol})^{-1}$
M_G Vapour flow rate, kg s^{-1}
M_L Liquid flow rate, kg s^{-1}
M_V Valve mass, kg
m Slope of equilibrium line (binary)
m $\Delta y/\Delta x$ defined variously by eqns. (10.2), (10.4), (10.6)

N Molar flux, kg-mol s^{-1} m^{-2}
N Number of completely mixed liquid pools
N_A Number of actual trays
N_G Number of binary vapour phase transfer units
N_L Number of binary liquid phase transfer units, eqn. (8.12)
N'_L Number of binary liquid phase transfer units, eqn. (8.22)
N_{OG} Number of overall binary vapour phase transfer units
N_p Total rate of drop generation per unit tray area, m^{-2} s^{-1}
N_T Number of theoretical trays
n Tray number from bottom of column

n Total number of bubble subgroups
n Number of holes in area A_b
n Parameter in eqn. (2.26)
n Parameter describing jet shape, eqn. (2.17)
n_1, n_2 Number of light and heavy valves, respectively

P Pressure, $N\,m^{-2}$
P_b Pressure within bubble, $N\,m^{-2}$
P_c Chamber pressure, $N\,m^{-2}$
P_I Excess pressure in bubble due to liquid inertia, $N\,m^{-2}$
P_s Pressure above liquid surface, $N\,m^{-2}$
ΔP_R Residual pressure drop, $N\,m^{-2}$
P_1^d Pressure at position 1 for dry tray, $N\,m^{-2}$
P_1^w Pressure at position 1 for wet tray, $N\,m^{-2}$
ΔP_{12} Pressure difference between 1 and 2, $N\,m^{-2}$
Pe Liquid Peclét number, $Q_L D/W h_{cl} De$ for two-dimensional model or $Q_L Z/W h_{cl} De$ for one-dimensional model
Pe_G Vapour Peclét number
p Hole pitch, m

Q_G Gas or vapour flow rate, $m^3\,s^{-1}$
Q_F Froth flow rate, $m^3\,s^{-1}$
Q_h Gas flow rate through hole, $m^3\,s^{-1}$
Q_L Liquid flow rate, $m^3\,s^{-1}$
Q_n Rate of heat loss from tray n to the surroundings, kW
q Heat flux, $kW\,m^{-2}$

R_h Hydraulic radius, m
Re_f Reynolds number for flowing two-phase mixture
Re_h Reynolds number for vapour flow through hole
r Bubble radius, m
r Ratio of N_1 to total flux
r_h Hole radius, m
r_m Manometer tube radius, m
r_n Fraction of heat lost by vapour from tray n

S Stabilisation index, $N\,m^{-1}$ or $N^2\,m^{-2}$
S Effective liquid flow path width for circular tray, m
SF System factor
Sc_G, Sc_L Schmidt number ($\mu_G/\rho_G D_G$ or $\mu_L/\rho_L D_L$)
s Distance of bubble centre from tray surface, m
s Parameter in drop size distribution, eqn. (2.33)
s Distance normal to curved wall, m
s' Parameter in drop size distribution, eqn. (2.31)

T Temperature, K
T_{Ln} Temperature of liquid leaving tray n, K

T^*_{Ln} Bubble point temperature of liquid leaving tray n, K
T_s Tray spacing, m
T_{Vn} Temperature of vapour leaving tray n, K
T^*_{Vn} Dew point temperature of vapour leaving tray n, K
t Time, s
t Tray thickness, m
t_G Mean residence time of vapour in dispersion, s
t_L Mean residence time of liquid on tray, s

u Gas velocity in jet, m s^{-1}
u_b Bubble rise velocity, m s^{-1}
u_{b0} Terminal rise velocity of isolated bubble, m s^{-1}
u_d Liquid velocity in downcomer on vapour-free basis, m s^{-1}
u_{dc} Critical value of u_d for downcomer choking, m s^{-1}
u_f Mean horizontal froth velocity, m s^{-1}
u_G Mean vapour velocity through dispersion, m s^{-1}
u_h Gas velocity through hole(s), m s^{-1}
u_L Mean liquid velocity across tray, m s^{-1}
u_{L1} Liquid velocity at position 1, m s^{-1}
u_s Superficial vapour velocity based on A_b, m s^{-1}
u'_s Superficial vapour velocity based on A_n, m s^{-1}
u_{smax} Maximum value of u_s, m s^{-1}
$(u_s)_{OBP}, (u_s)_{CBP}$ Value of u_s at open and closed balance points, m s^{-1}

V_b Bubble volume, m^3
V_c Chamber volume, m^3
V_i Liquid molar volume at normal boiling point, cm^3 (g-mol)$^{-1}$
v Drop velocity, m s^{-1}
v_o Fraction of open valves
v_p Drop projection velocity, m s^{-1}

W Weir length, m
We Weber number
w, w_1 $w = w'/D$, $w_1 = W/2D$
w' Distance from centre line measured parallel to weir, m

X Mole fraction in pseudo-binary liquid mixture
x Mole fraction in liquid
x_{in} Tracer concentration at tracer injection point
x_0 Tracer concentration in liquid entering tray
x_R Mole fraction in liquid leaving reboiler, Fig. 9.5
x^L Mole fraction of low surface tension component
x_+ Value of x immediately downstream of liquid entry
x^*_{en-1} Mole fraction of liquid in equilibrium with vapour entering tray n
$\bar{x}_n$ Mean mole fraction in liquid leaving tray n via downcomer

x_n^* Mole fraction of liquid in equilibrium with mean vapour concentration leaving tray n
x_n' Mole fraction in liquid weeping or entrained from a point on tray n
$\bar{x}_n'$ Mean mole fraction in liquid weeping or entrained from tray n averaged over the tray

Y Mole fraction in pseudo-binary vapour mixture
$\bar{Y}_n$ Defined by eqn. (9.24) for entrainment, eqn. (9.33) for weeping
y Depth below free surface, m
y Parameter in eqns. (2.31), (2.33)
y Mole fraction in vapour
y_n Value of y in vapour leaving a point on tray n
$\bar{y}_n$ Mean value of y in vapour leaving tray n
y^* Mole fraction in vapour which is in equilibrium with liquid
y_n^* Mole fraction in vapour in equilibrium with mean concentration of liquid leaving tray n via the downcomer

Z Parameter in eqn. (4.19)
Z Liquid flow path length, m
Z_e Length of exit calming zone, m
Z_F Film thickness, m
z Distance normal to interface, Chapter 8, m
z $z = z'/D$ in two-dimensional model; $z = z'/Z$ in one-dimensional model
z' Distance from inlet weir, m
z_1 Value of z at exit weir
z_{in} Value of z at tracer injection point

α Volume of liquid per unit volume of two-phase dispersion (liquid holdup fraction)
α Relative volatility, eqn. (7.23)
α Similarity ratio, Chapter 9
$\alpha_d, \bar{\alpha}_d$ Local and mean liquid volume fraction in the downcomer
α_e Effective liquid volume fraction defined by eqn. (3.24)
$\alpha(i)$ Local liquid volume fraction at height $h(i)$
α_R Liquid volume fraction in froth flowing under downcomer
$\alpha_{OG}, \alpha_G, \alpha_L$ Parameters in eqns. (10.7)
β_0 Fractional weeping rate, L_w/L_0
γ Parameter in eqns. (10.10)
γ_{ni}^L Activity coefficient of component i in liquid leaving tray n
ε Volume of gas or vapour per unit volume of two-phase dispersion (gas holdup fraction)
ε_w Value of ε in froth flowing over weir
ε_1' Local volume fraction of small bubbles
η $\varepsilon/(1-\varepsilon)$, eqn. (3.10)
θ Contact angle
$\theta(ij)$ Residence time of drops j in height increment at height $h(i)$, s

λ Stripping factor or ratio of slope of equilibrium line to slope of operating line, mG/L
λ_0 mG/L_0
μ_G, μ_L Vapour and liquid viscosity, $N\,s\,m^{-2}$
v Volume fraction of drops having diameter less than y, eqn. (2.31)
ξ Orifice coefficient
ξ' Modified orifice coefficient, eqn. (4.18)
ξ_0 Orifice coefficient as $\phi \to 0$, Fig. 4.2
ξ_{VC} Orifice coefficient for valves resting on tray deck
ξ_{VD} Orifice coefficient for valve tray deck
ξ_{Vi} Orifice coefficient for combination of light and heavy valves
ξ_{VO} Orifice coefficient for open valves
ρ_F Density of two-phase dispersion, $kg\,m^{-3}$
ρ_G, ρ_L Vapour and liquid density, $kg\,m^{-3}$
ρ'_G, ρ'_L Vapour and liquid density, kg-mol m^{-3}
ρ_{H_2O} Density of water, $kg\,m^{-3}$
σ Surface tension, $N\,m^{-1}$
$\sigma^+, \sigma^-, \sigma^0$ Surface tension positive, negative or neutral system
ϕ Fractional perforated tray area (hole area/bubbling area); also called fractional free area
ϕ_A Association factor
ψ Defined by eqn. (3.22*a*), m
ψ Parameter in eqn. (10.5)

$\mathcal{N}_{OG}, \mathcal{N}_G, \mathcal{N}'_L$ Number of overall vapour phase, vapour phase and liquid phase ternary transfer units

Subscripts

b In bulk phase
G In gas or vapour phase
i For bubble subgroup *i*
i For component *i*
i At interface
ij For the binary *ij*
j For the jet
j For component *j*
L In liquid phase
n Leaving tray *n*
V In vapour phase
1 or 2 For component 1 or 2, respectively *or* for small or large bubbles, respectively

Superscripts

e Exit stream from ideal tray
0 Local value at the exit weir

— Mean value
$\dot{}, \ddot{}$ First and second derivatives with respect to time
$''$ Entering value used in ideal tray definition, Fig. 7.1

1

Some general considerations

1.1 Introduction

Distillation has a long history. Reputedly it was the Chinese who discovered it during the middle of the Chou dynasty. Thereafter, the production of distilled liquors, the so-called liquids of the gods, followed the progress of civilisation. First India, then Arabia, the secret reached Britain before AD 500 as the production of mead. Surprisingly, it took a further millennium before whisky was first distilled in Scotland about AD 1500. The full history of distillation has been meticulously chronicled by Forbes (1948) and that specific to North America by Carr (1972).

Although alcoholic beverage production retains its importance for many, distillation plays a far greater role in human affairs today, for it is now the dominant separation process used in the petroleum and chemical industries. It has achieved this dominance, and seems likely to retain it, despite its apparently wasteful use of energy. Alternatives to distillation, such as solvent extraction, adsorption or membranes, can be more energy efficient, but they often have more than offsetting higher investment costs. As a result, distillation retains its advantage, particularly in large-scale applications. Because of its massive scale of operation, even small improvements in distillation can have significant impact, and Zuiderweg (1973) has estimated that two billion dollars in column investment costs alone were saved between 1950 and 1970 by research and development.

Turning now to distillation trays, the theme of this book, there is some evidence that a rudimentary form of sieve tray was employed by the Greeks in about the second century AD. However, it was in response to a competition sponsored by Napolean Bonaparte that continuous distillation using bubble cap trays was discovered by Cellier-Blumenthal in 1813. Sieve trays, as we now know them, were apparently first employed in the Coffey still in 1830. Fair (1983) has given a comprehensive review of the historical development of column internals.

The two dominant classes of column internals used today are trays and packing. Packing has a pressure drop which is about one-fifth that of trays. Consequently, it is often the preferred choice in cases where pressure drop is an overriding consideration such as vacuum distillation or where vapour recompression (heat pumping) is used. Sometimes a lower temperature heating medium can be employed in the reboiler using packing, but this is not always significant. An example of such a case is where quench water or steam is in surplus and at an appropriate temperature. Recent developments have made packed column scale-up less uncertain, but the more positive directed flow of each phase in a trayed column (at the expense of pressure drop) makes hydraulic and mass transfer behaviour more predictable for trays than for packing. Factors such as those shown in Table 1.1, and elsewhere (Thibodeaux & Murrill 1966, Fair 1970, Billet, Conrad & Grubb 1969), have to be considered for each application. It has been estimated that currently about 90% of installed distillation columns contain trays (Krummrich 1984).

1.2 Tray types

Factors which influence the selection of tray type include capacity, efficiency, turndown, pressure drop, fouling resistance, cost and, not least, tradition. Figs. 1.1 and 1.2 show simple representations of most of the tray types in common use classified by deck design and by flow path arrangement. A very large number of tray types are possible by combining different decks and flow path arrangements. Table 1.2 summarises points to

Table 1.1. *Trays or packing – some factors to consider*

	Trays	Random packing	Structured packing
Effect of scale-up on HETP[a]	Predictable	Difficult to predict	Predictable
Pressure drop	High	Low	Low
Established design techniques	Yes	Only for capacity, not for HETP	
Cost	Low	Low–medium	High
Suitability for fouling service	Yes	No	No
Feed point flexibility	Easy	Difficult	Difficult

[a] HETP – height of an equivalent theoretical plate

consider when selecting between the various options and indicates sources of more detailed information on each device. The tray patent literature illustrates the wide variety of tray types which have been proposed and it has been summarised by Jamal (1981). The great majority of trays currently being installed are either sieve or valve trays and this is reflected in the topics covered in subsequent chapters.

1.3 Classifying distillation systems

1.3.1 *Variation of physical properties with flow parameter*

The flow parameter (FP), defined by eqn. (1.1), is a useful dimensionless group which is frequently used in tray hydraulics correlations:

$$FP = \frac{M_L}{M_G} \cdot \left(\frac{\rho_G}{\rho_L}\right)^{0.5} \tag{1.1}$$

Except for easy separations of high relative volatility, the reflux ratio in distillation tends to be large such that M_G and M_L are not very different.

Fig. 1.1. Some styles of deck design in trayed columns.

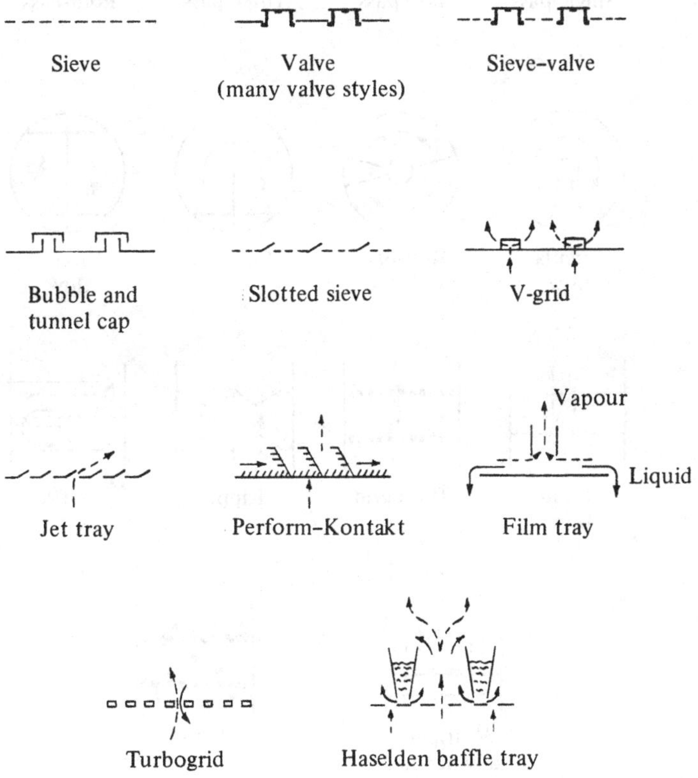

Consequently, it is often a reasonable approximation to assume that

$$FP \approx \left(\frac{\rho_G}{\rho_L}\right)^{0.5} \tag{1.2}$$

Porter & Jenkins (1979) pointed out that, for typical distillation systems, changes in physical properties can often be correlated against each other. Thus, physical properties can be correlated against (ρ_G/ρ_L) or against FP using eqn. (1.2). Their suggested correlation is shown in Fig. 1.3. The correlation holds approximately for any combination of temperature and pressure providing they correspond to saturation conditions. A further useful approximation is achieved by noting that distillation is often carried out between 50 and 150°C. These limits allow condensation of overhead vapour against cooling water and avoid thermal degradation of bottom

Fig. 1.2. Some flow-path arrangements used in trayed columns.

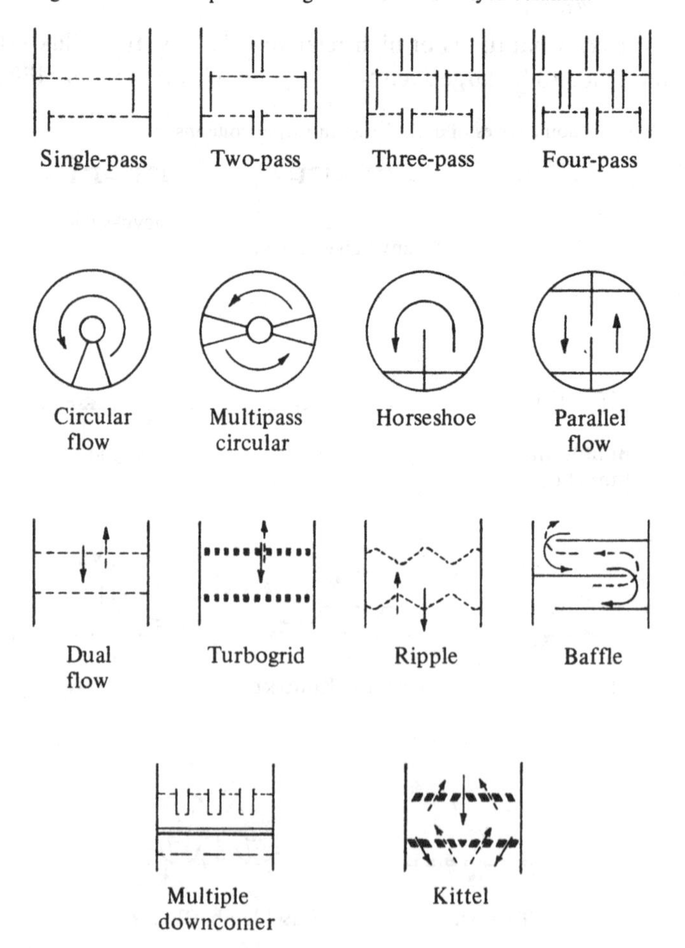

Table 1.2. *Some factors in choosing tray type*

Type	Comments	References
Sieve	Low cost, versatile, non-proprietary	
Valve	Lower turndown, 5–10% higher cost than sieve	Glitsch Inc. (1974) Koch Engineering Co. (1982) Nutter Engineering Co. (1976)
Sieve–valve	Lower turndown, higher efficiency than sieve	Billet *et al.* (1969)
Bubble cap	For extremely low turndown, high cost	Bolles (1956)
Slotted sieve	High efficiency, low pressure drop	Smith & Delnicki (1975)
V-grid	Less entrainment and weeping than sieve	Nutter (1971)
Jet	High liquid capacity – prone to liquid blow-off	Forgrieve (1960) Kirsten & Van Winkle (1970)
Perform–Kontakt	High vapour and liquid capacity	Raskop (1974)
Film	Low pressure drop	Leva (1972)
2–5 pass	For high liquid loads, prone to maldistribution, avoid odd number of passes	Bolles (1976*b*)
Multiple downcomer	For high liquid loads, high capacity	Union Carbide Corp. (1970) Delnicki & Wagner (1970)
Circular flow/horseshoe	Low liquid loads – Lewis's case 2	Bolles (1963)
Multipass circular	Higher liquid loads – Lewis's case 2	Ying *et al.* (1984)
Parallel flow	Maximum bubbling area – Lewis's case 2	Smith & Delnicki (1975)
Dual flow/turbogrid	Fouling services, poor turndown, low efficiency	Rylek & Standart (1964)
Ripple	Fouling services, handles solids	Hutchinson & Baddour (1956)
Baffle	Severe fouling or polymerisation service	Lemieux (1983)
Haselden's baffle tray	High capacity, high cost	Haselden & Witwit (1981)
Kittel	Low cost	Stanislas & Smith (1960)

product in the reboiler. Under this restriction, typical operating pressures can be included on Fig. 1.3 as shown by the broken line. Doig (1971) has also given a detailed discussion of the factors to consider in choosing the operating pressure for distillation. Clearly, inclusion of pressure on Fig. 1.3 is inappropriate when refrigeration is used, as for demethanisers or for air distillation, nor does it apply if overhead vapour recompression is used. Furthermore, it is too inaccurate to be used for design. Nevertheless it provides useful orientation. For example, it indicates that high-pressure distillation, corresponding to low-molecular-weight systems, is associated with low surface tension. This has important implications for column flooding and is discussed in Chapter 5.

Fig. 1.3. Typical variation of physical properties with flow parameter in distillation (Porter & Jenkins 1979). Ordinate multiplying factors ( ): for gas viscosity, $N\,s\,m^{-2}$, multiply ordinate by 10^{-8}; liquid viscosity, $N\,s\,m^{-2}$ ($\times 10^{-6}$); gas self diffusion coefficient, D_G, $m^2\,s^{-1}$ ($\times 10^{-7}$); liquid self diffusion coefficient, D_L, $m^2\,s^{-1}$ ($\times 10^{-9}$); Schmidt numbers ($\times 10^{-1}$); pressure, bar ($\times 10^{-2}$); gas density, $kg\,m^{-3}$ ($\times 10^{-1}$); liquid density, $kg\,m^{-3}$ ($\times 10$); surface tension, $N\,m^{-1}$ ($\times 10^{-3}$).

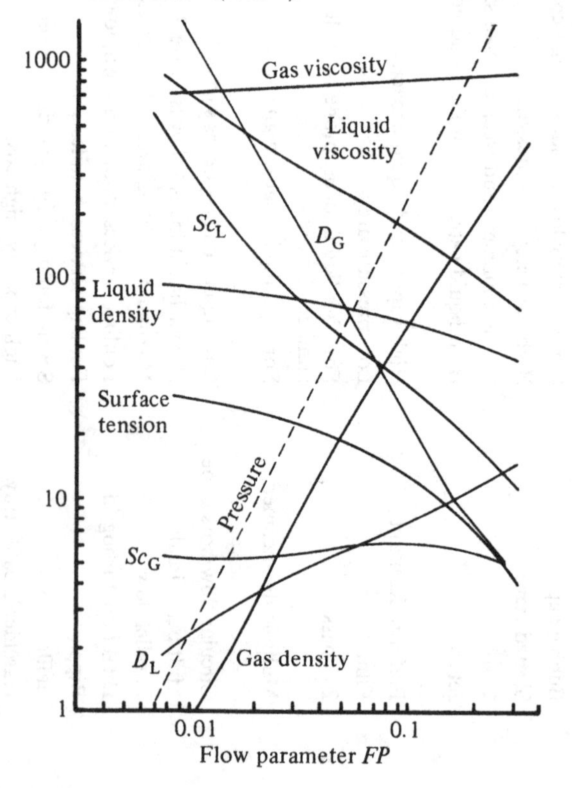

1.3.2 *Change of flow regime with flow parameter*

The flow regime diagram of Fig. 1.4 indicates how the flow regime on the tray typically changes with flow parameter. The precise location of the transition lines between regimes is open to dispute and can be established using the correlations outlined in subsequent chapters. The transitions also depend on the details of the tray design. However, Fig. 1.4 in combination with Fig. 1.3 does indicate that typically:

- vacuum distillation can result in spray regime operation;
- atmospheric distillation typically involves operation in the froth regime;
- high-pressure distillation is usually associated with the emulsion regime.

The characteristics of each flow regime are discussed in Chapter 2.

Again it must be emphasised that, useful though these generalised figures are, they only give a rough indication of trends and each case must be considered in detail. As an example, Figs. 1.3 and 1.4 imply that the separation of ethylbenzene-styrene, a typical vacuum distillation system, is associated with spray regime operation. In fact, this system is dominated by the need to minimise pressure drop so as to limit polymer formation in the reboiler. As a result, rather lower than normal superficial vapour velocities are used and typically the trays operate in the froth regime; see Fig. 1.5 (Lockett, Plaka & Ahmed 1984).

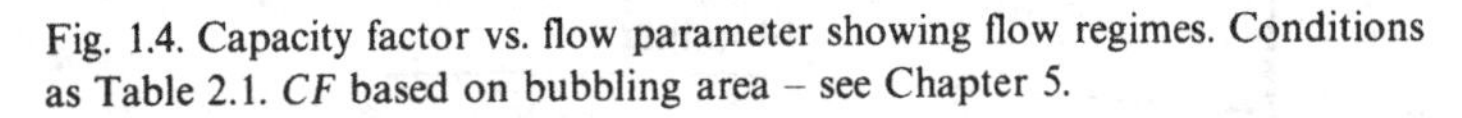

Fig. 1.4. Capacity factor vs. flow parameter showing flow regimes. Conditions as Table 2.1. *CF* based on bubbling area – see Chapter 5.

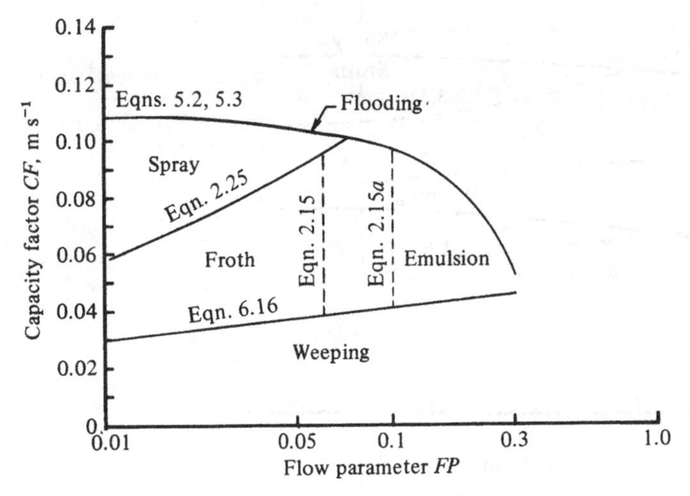

1.3.3 *Variation of column diameter and number of passes with loading*

Useful orientation can be obtained from the total flows chart developed by Porter & Jenkins (1979) shown in Fig. 1.6. It is based on unpublished charts prepared by the author. Fig. 1.6 was constructed using flooding data released by Fractionation Research, Inc. (Sakata & Yanagi 1979), and a constant downcomer liquid velocity of 0.19 m s^{-1}. Also included on Fig. 1.6 are lines corresponding to some typical pressures for distillation at total reflux plotted using Fig. 1.3. The chart quantifies common experience that high pressures are associated with multipass trays, whereas low-pressure distillation is usually carried out using single-pass trays. A more detailed discussion can be found in Section 5.1.3.

1.4 An outline design procedure

The basic steps involved in designing distillation trays have been well documented both for new columns and for retrays of existing columns (Billet 1979, Backhurst & Harker 1973, Bolles 1956, 1963, Chase 1967, Economopoulos 1978, Fair 1963, Frank 1977, Glitsch, Inc., 1974, Kister 1980, Koch & Kuzniar 1966, Koch Engineering Co. 1982, Neretnieks 1970, Nutter Engineering Co. 1976, Raper *et al.* 1977*b*, Sewell 1975, Stichlmair 1978). There is considerable creative skill and experience required to arrive

Fig. 1.5. Typical capacity factor vs. flow parameter variation for 7.0 m-diameter single-pass sieve tray in vacuum distillation (Lockett, Plaka & Ahmed 1984). ---- Tray pressure drop (mm Hg).

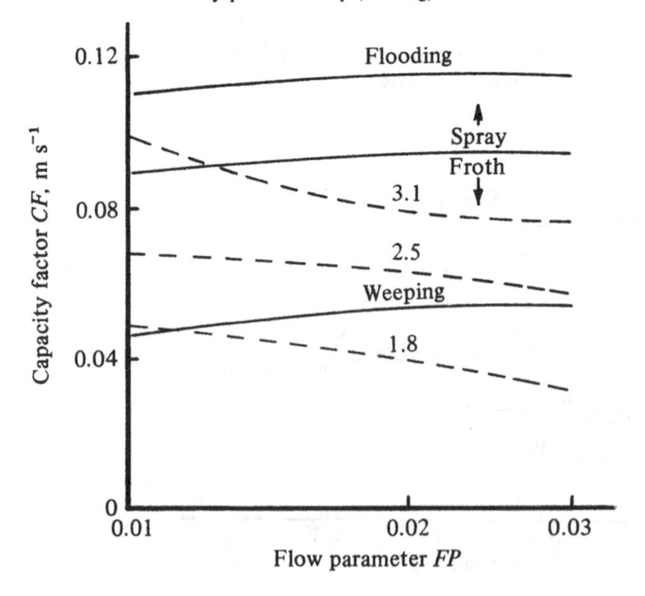

at a safe design but one which is also cost effective by incorporating a minimum of overdesign. The reason for this is that most of the variables involved interact in a complex way.

Set out below is an outline of a typical design procedure for sieve trays with some discussion of the tradeoffs involved. A similar procedure applies to valve trays. Reference is made to subsequent sections where more details are given.

(1) Fix the number of passes, tray spacing and hole diameter based on experience in similar applications. As a guide, use Fig. 1.6 to give the number of passes, and choose an initial tray spacing of 0.61 m. There are two schools of thought about hole diameter. In one, a hole diameter of 12.7 mm is nearly always used. A better approach is to use a hole diameter of 4.8–6.4 mm unless fouling or corrosion are likely to be excessive. Smaller holes give a higher vapour capacity and can allow increased turndown, although perforation costs are slightly higher.

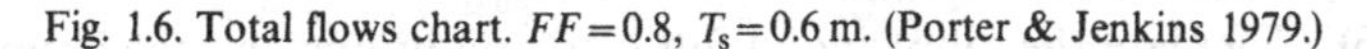

Fig. 1.6. Total flows chart. $FF = 0.8$, $T_s = 0.6$ m. (Porter & Jenkins 1979.)

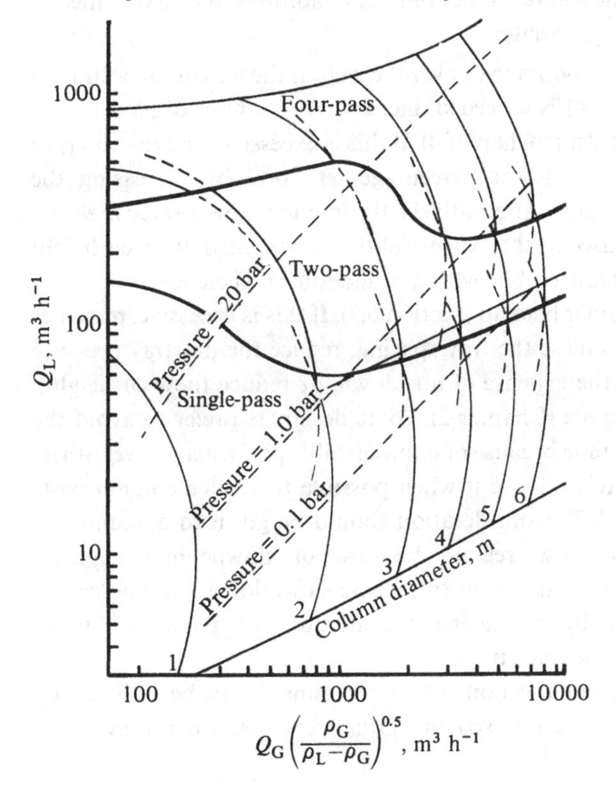

(2) Fix the turndown required (Chapter 6). A typical turndown for a sieve tray is 60–70% of full load flow rates. Sieve trays can be designed for far lower turndown, but at the expense of pressure drop and often of increased tray spacing. Since the provision of turndown is expensive, it is important to determine at the outset whether the turndown asked for is really required.
(3) Fix the exit weir height. A typical value is 50 mm, with 0–25 mm used in vacuum distillation and 100 mm in absorbers and strippers. Increasing the weir height increases the tray efficiency but at the expense of pressure drop.
(4) Determine the bubbling area and downcomer area from capacity correlations (Chapter 5).
(5) At turndown conditions, determine the fractional perforated tray area to ensure tray stability and minimise weeping at an acceptable level (Chapter 6 and Section 9.11). Since the weep point and weeping correlations involve the clear liquid height, which in turn depends on the fractional perforated area (Chapter 3), an iterative procedure is required.
(6) Calculate the following at full load conditions and take remedial action as appropriate:
   (*a*) Maximum liquid load over the weir. If the maximum weir load (Section 5.4) is exceeded, increase the number of passes.
   (*b*) Pressure drop (Chapter 4). If this is excessive, one remedy is to design at a lower percentage jet flood by increasing the bubbling area. Alternatively, the fractional perforated area can be increased at the expense of turndown. Also, the weir height can be reduced but with a reduction in efficiency.
   (*c*) Downcomer backup (Section 5.3). If this is excessive, remedies are to increase the tray spacing, reduce the dry tray pressure drop at the expense of turndown, or reduce the weir height.
   (*d*) Flow regime (Chapter 2). Some designers prefer to avoid the spray regime because of unpredictable performance, yet others deliberately choose it when possible to reduce column costs (Section 8.7). Consideration should be given to avoiding the emulsion flow regime because of downcomer capacity limitations and vapour entrainment (Section 5.3.2). This can be achieved by increasing the number of passes or using multidowncomer trays.
   (*e*) Entrainment (Section 5.2). Entrainment can be reduced by increasing the tray spacing. Other remedies are also available

to reduce entrainment which differ depending whether the tray operates in the froth or spray regimes.

(7) Estimate the tray efficiency. If this proves to be unacceptably low, options include increasing the weir height at the expense of pressure drop, downcomer backup and turndown. Also the bubbling area can be increased to increase vapour–liquid contact time.

(8) The above steps are usually repeated for different initial tray spacings until an acceptable minimum cost design has been achieved. Cost includes not only the trays but also the column shell, foundations, erection, piping and insulation. Additional factors which have to be considered are the maximum allowable column height (50–90 m depending on local conditions) and the height-to-diameter ratio. If the latter exceeds 25, problems can occur in supporting the column. A minimum tray spacing of perhaps 450 mm may also sometimes be specified for column access.

In practice the tray design procedure described above is only part of an overall column design in which the tradeoff between investment and operating costs is considered as a function of reflux ratio. However, it often turns out that the optimum reflux ratio is approximately 20% above the minimum value and is fairly insensitive to variations in column investment cost. In this case the incentive is simply to minimise the installed cost of the column (Haselden 1975).

2

Bubbles, froth, spray and foam

2.1 Introduction

This chapter deals with the flow regimes found on trays and the transitions between them. Classification by flow regime is an approach which has been used successfully in the study of two-phase flow in pipes. The major flow regimes on trays are:

Spray regime – This regime is gas phase continuous and the liquid is projected up by the gas jets to form small drops. It has been likened to an inverted rainstorm and is favoured by high gas momentum and low liquid depths; see Fig. 2.1*a*.

Froth regime – Here the situation is reversed and the liquid is the continuous phase. Gas passes through the liquid as jets and bubbles of ill-defined and rapidly changing shape. There is a range of bubble sizes present; see Fig. 2.1*b*.

Emulsion regime – Small gas bubbles are 'emulsified' in the liquid. The regime is favoured by a high horizontal liquid momentum compared with the vertical gas momentum; see Fig. 2.1*c*.

Bubble regime – Bubbles rise in swarms through a fairly quiescent liquid. This regime exists only a very low gas rates; see Fig. 2.1*d*.

Foam regime – Foam tends to result when bubble coalescence is hindered. Various types of foam are possible. Cellular foam, composed of large dodecahedral bubbles, can form at low gas and liquid rates. It has a tendency to occur with aqueous systems in columns of small diameter; see Fig. 2.1*e*. More common mobile foams are formed when bubble coalescence is inhibited in a froth and can be particularly troublesome in the emulsion regime.

Why is the study of flow regimes important? One reason is that tray hydrodynamic behaviour depends on the regime. Different correlations are required in each regime for such things as dispersion density and entrainment. As an example, the effect of regime on dispersion density profile (measured by γ-ray absorption) is shown in Fig. 2.2. Another reason is that some regimes are worth avoiding. For example, foaming is exacerbated in the emulsion regime and entrainment increases rapidly in the spray regime. A third reason is that there is some evidence that tray efficiency can be improved by designing the tray to operate in an appropriate regime (Section 8.7).

Fig. 2.1. Representation of flow regimes.

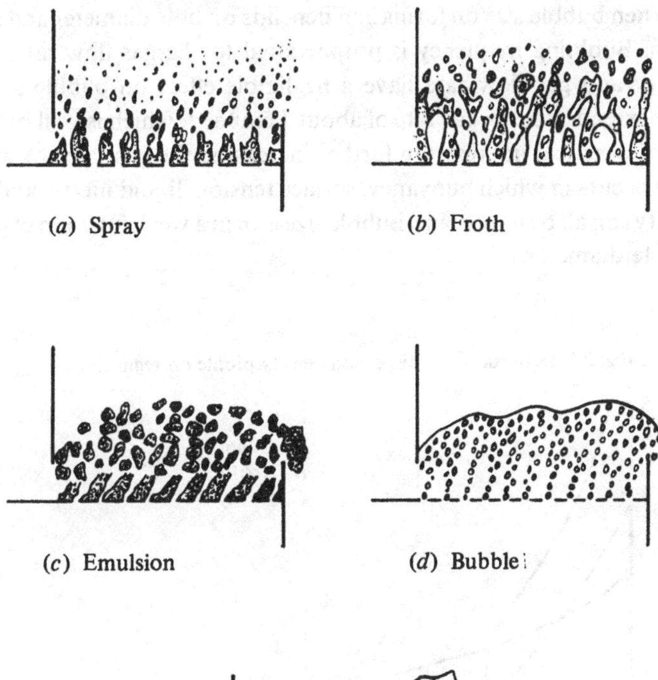

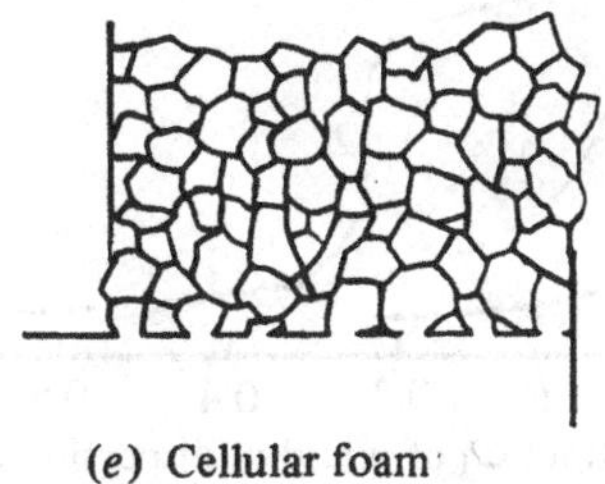

(*e*) Cellular foam

2.2 Bubbling from single holes

Many studies of bubbling from single holes have been reported in an effort to elucidate the mechanism of bubbling on multihole sieve trays. For the most part the results have been disappointing. Models have been developed only for fairly simple ideal bubbling at low gas rates and have not yet been extended satisfactorily to multihole trays at industrial flow rates.

To help place the single hole studies in context, Table 2.1 shows the typical range of hydrodynamic variables encountered in distillation using sieve trays, shown as a function of flow parameter. Note the high hole Reynolds numbers and the wide range of vapour flow rates per hole.

2.2.1 *Qualitative description of bubbling from single holes*

Here we consider bubbling into a deep liquid pool.

Static regime. The static or constant volume regime occurs at very low gas rates when bubble size on formation depends on hole diameter and surface tension. Bubbling frequency is proportional to the gas flow rate. Liquid viscosity and gas flow rate have a negligible effect on bubble size. The regime extends up to a gas rate of about 1 $cm^3 s^{-1}$, which is well below the range of practical interest. On further increasing the gas rate, a transition regime occurs in which buoyancy, surface tension, liquid inertia and liquid viscosity can all be important. Bubble size is then a weak function of gas rate and hole diameter.

Fig. 2.2. Dependence of dispersion density profile on regime.

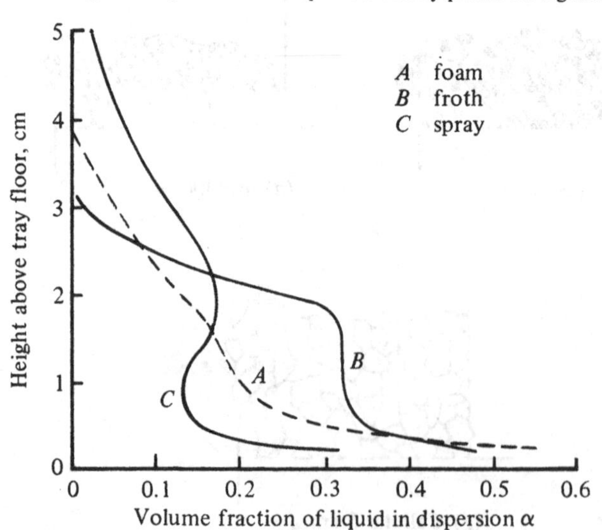

Table 2.1. *Typical values of hydrodynamic parameters in distillation*

FP	ρ_G (kg m^{-3}) (1)	ρ_L (kg m^{-3}) (1)	μ_G (N s m^{-2}) (1)	σ (N m^{-1}) (1)	u_s (m s^{-1}) (2)	Re_h (3)	Re_h (4)	Q_h (cm^3 s^{-1}) (3)	Q_h (cm^3 s^{-1}) (4)	Q_L/W (m^3 m^{-1} s^{-1})	h_{cl} (m) (3)(5)
0.02	0.30	900	7.5×10^{-6}	25×10^{-3}	5.95	30 000	11 000	7500	1100	3.2×10^{-3}	0.025
0.05	2.0	800	8.0×10^{-6}	20×10^{-3}	2.08	66 000	25 000	2600	380	8.5×10^{-3}	0.032
0.10	6.0	600	8.5×10^{-6}	15×10^{-3}	0.98	88 000	33 000	1200	180	1.6×10^{-2}	0.038
0.30	40	450	9.0×10^{-6}	4.5×10^{-3}	0.17	96 000	36 000	220	31	2.5×10^{-2}	0.049

Column diameter 2.0 m
Single-pass sieve tray
$W/D=0.75$, $A_b=2.44\,m^2$, $A_d=0.35\,m^2$, $T_s=0.61$ m, $h_w=0.05$ m, $\phi=0.1$
Total reflux
Note for 60 deg. triangular pitch perforations:

$$\text{hole pitch } p=\left[\frac{\pi}{2\sqrt{3}}\cdot\frac{d_h^2}{\phi}\right]^{0.5}$$

Notes:
(1) From Fig. 1.3
(2) From eqns. (5.1)–(5.3):
u_s based on bubbling area A_b
flow rates reduce on turndown
(3) $d_h=0.0127$ m
(4) $d_h=0.0048$ m
(5) From eqn. (3.26)

Dynamic regime. At gas rates above about 15 $cm^3 s^{-1}$ the bubble frequency tends very approximately to be about 20 s^{-1}. The dominant forces acting on the bubbles are buoyancy, liquid and gas inertia, and sometimes liquid viscosity. This regime includes the lower end of the practical range of interest. Fig. 2.3 shows the types of bubbling obtained at atmospheric pressure and Fig. 2.4 shows the reported effect of gas inertia as determined

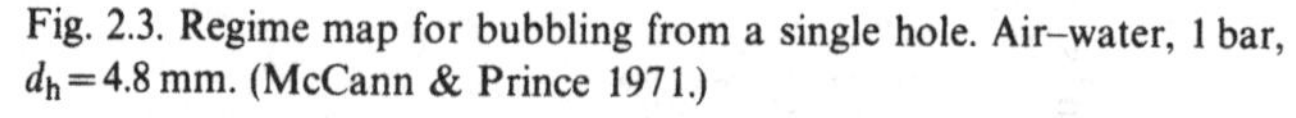

Fig. 2.3. Regime map for bubbling from a single hole. Air–water, 1 bar, d_h=4.8 mm. (McCann & Prince 1971.)

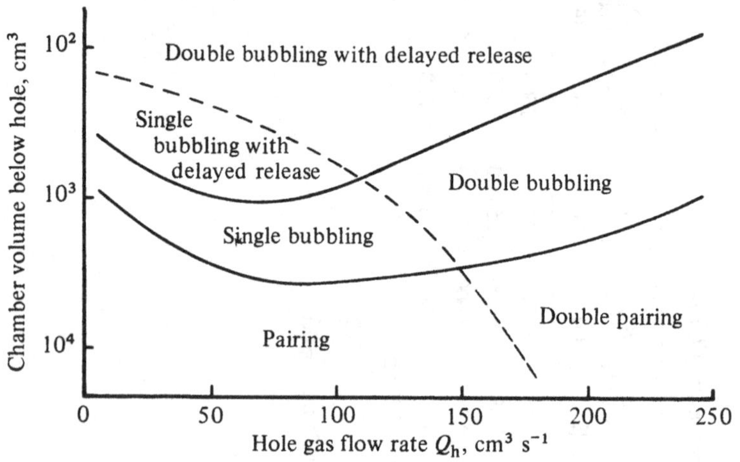

Fig. 2.4. Regime map for bubbling from a single hole at high pressures. CO_2–H_2O, V_c=375 cm^3, d_h=4.8 mm. (LaNauze & Harris 1974.)

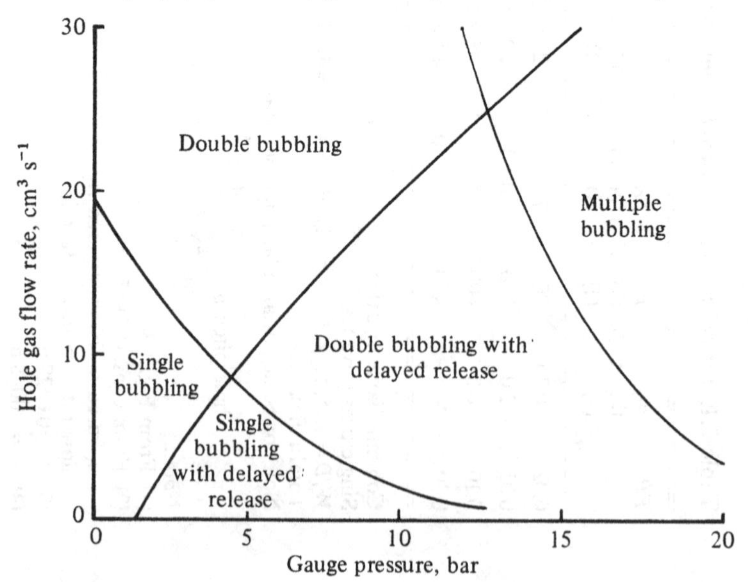

by pressure. The different types of bubbling shown in these figures are discussed below.

Single ideal bubbling, in which the bubbles are spherical during their formation and detachment, occurs only at low gas rates. The stages of bubble formation for single ideal bubbling, Fig. 2.5, have been described by Kupferberg & Jameson (1969) as follows:

(*a*) Growing stage: Bubble growth starts when the pressure below the plate is sufficient to overcome surface tension. The inertia of the liquid surrounding the bubble is initially large and the bubble grows slowly. For small chamber volumes below the plate, the gas enters the chamber faster than it leaves and the chamber pressure initially increases. As the bubble grows, liquid inertia is less significant and the pressure in the bubble falls since it corresponds to the hydrostatic pressure at its centre. Consequently, the rate of bubble growth increases and the chamber pressure tends to decrease.

(*b*) Elongating stage: The bubble continues to grow and accelerate while still connected to the hole by a small tail. Eventually the tail breaks and the bubble detaches. Chamber pressure continues to fall during this stage.

(*c*) Waiting stage: The chamber pressure builds up to a value sufficient to initiate the growth of a new bubble. If the pressure in the liquid behind the bubble at detachment is greater than the chamber pressure, weeping of liquid is possible during this stage.

The other types of bubbling involving multiple bubble formation shown in Figs. 2.3 and 2.4 have been described by McCann & Prince (1969, 1971).

Pairing occurs when the pressure behind the first bubble is less than the chamber pressure, which happens with large chamber volumes. A second small bubble is formed which joins and coalesces with the first bubble. The latter can continue to grow while fed with gas through the tube connecting it to the hole. *Double bubbling* occurs when the bubble frequency increases,

Fig. 2.5. Stages in ideal bubbling.

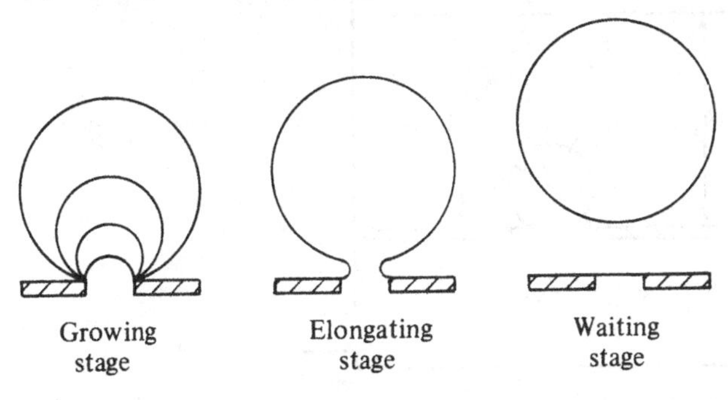

either by a reduction in chamber volume or by an increase in gas flow rate, and the wake of the first bubble affects the next one. The latter is deformed into an ellipse, is drawn up into the first bubble, and after coalescence the two bubbles rise as one.

For small chamber volumes *delayed bubble release* is possible when the chamber pressure falls to such a low value during the formation cycle that flow into the bubble ceases. There is a latent period while gas flows into the chamber and its pressure rises. Eventually the bubble resumes growth and subsequently detaches. Delayed release can occur with single or double bubbling. *Double pairing* occurs when pairing and double bubbling occur at the same time. It is unstable and two to four bubbles are formed randomly.

Bubble coalescence takes place closer and closer to the hole as gas flow rate or gas density is increased, leading to *multiple bubbling* or chain bubbling (LaNauze & Harris 1974, Wraith 1971). When photographs are taken using a high-speed camera, it can be seen that what appears to the naked eye as a pulsating jet is in fact multiple bubbling. A transition line between double bubbling and multiple bubbling has been tentatively suggested at high pressures (Fig. 2.4) but at atmospheric pressure the transition has not been identified.

2.2.2 *Simple model for ideal bubbling from a single hole*

General equation. A number of similar mathematical models have been developed which are strictly only applicable to single ideal bubbling and to pairing (Davidson & Schuler 1960*a*, *b*, Kupferberg & Jameson 1969, McCann & Prince 1969, Satyanarayan *et al.* 1969, Ramakrishnan *et al.* 1969, LaNauze & Harris 1972, 1974, Wraith 1971). Other models have been reviewed by Clift *et al.* (1978). A brief generalised treatment is given here based on Fig. 2.6. It is assumed that the bubble remains spherical

Fig. 2.6. Nomenclature for bubble formation model.

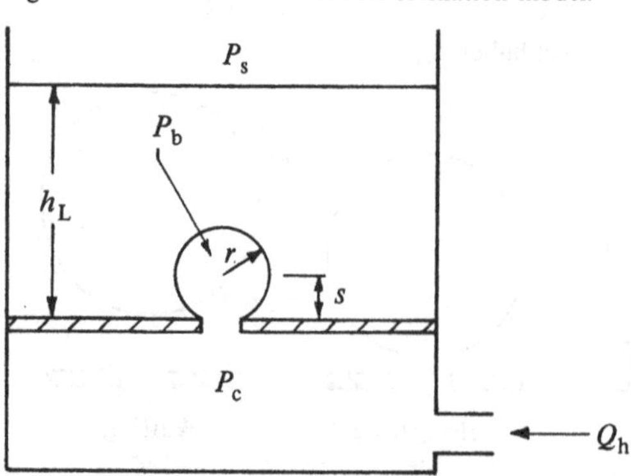

throughout its formation and that it is unaffected by the wake of the previous bubble. A force balance on the bubble and its surrounding liquid gives

$$V_b(\rho_L-\rho_G)g+\frac{\rho_G \dot{V}_b^2}{A_h}+F_R=\left(\rho_G+\frac{11}{16}\rho_L\right)\frac{d}{dt}(V_b\dot{s})+6\pi\mu_L r\dot{s}+\pi d_h\sigma\cos\theta$$

(i) (ii) (iii) (iv) (v) (vi)

(2.1)

The terms represent (i) buoyancy, (ii) rate of change (with time) of gas momentum entering the bubble, (iii) reaction of the plate, (iv) rate of change of bubble momentum, where the bubble virtual mass is taken as that of a sphere moving perpendicular to a wall in an inviscid fluid, (v) viscous drag, (vi) surface tension force (zero after detachment). Some authors (McCann & Prince 1969) ignore the influence of the plate on the virtual mass, when $\frac{11}{16}$ becomes 0.5. Kupferberg & Jameson (1969) derived an equivalent equation to eqn. (2.1) from first principles by considering the pressure field around a spherical bubble situated close to a plate.

Static regime. In the static regime, terms (ii), (iv) and (v) are negligible and at detachment, when $F_R=0$, the remaining terms give

$$V_b=\frac{\pi d_h\sigma\cos\theta}{(\rho_L-\rho_G)g} \tag{2.2}$$

Dynamic regime – constant flow. In the dynamic regime, a simple solution is possible (Davidson & Schuler 1960*a*) when terms (ii), (v) and (vi) are negligible and the hole gas flow rate remains constant at Q_h. As an idealisation it is assumed that the bubble grows from a point source ($F_R=0$), and although spherical $s<r$ until detachment. Hence

$$V_b g=\frac{11}{16}\frac{d}{dt}(V_b\dot{s}) \quad \rho_G \ll \rho_L$$

Since $V_b=Q_h t$, integrating twice and using the initial conditions $t=0$, $s=\dot{s}=0$, gives $s=4gt^2/11$. At detachment,

$$s=r=\left(\frac{3V_b}{4\pi}\right)^{1/3}=\left(\frac{3Q_h t}{4\pi}\right)^{1/3}$$

Eliminating s gives

$$V_b=Q_h t=1.378\frac{Q_h^{6/5}}{g^{3/5}} \tag{2.3}$$

Hence,

$$\text{bubbling frequency}=0.73g^{0.6}Q_h^{-0.2} \tag{2.4}$$

The assumption of constant gas flow above leads to the definition of two limiting cases:

(*a*) Constant pressure regime. For large chamber volumes below the plate, the pressure fluctuations in the chamber caused by bubble formation and release are small. The chamber pressure therefore effectively remains constant. Consequently the gas flow through the hole varies in response to changes in pressure within the bubble.

(*b*) Constant flow regime. For small chamber volumes, as the bubble forms from gas stored in the chamber, the pressure in the chamber falls in response to the bubble's demand. As a result, the pressure drop across the hole tends to remain constant and the flow rate through the hole also remains constant. Large values of gas flow rate or hole resistance lead to large hole pressure drops, and therefore changes in chamber or bubble pressure have only a small effect on pressure drop. Consequently, these conditions also tend towards the constant flow regime.

In practice the real situation usually falls somewhere between these two limiting cases.

Dynamic regime – general solution. When constant flow cannot be assumed, it is necessary to consider the pressures within the bubble and chamber. For the hole

$$P_c - P_b = 0.5\,\xi \rho_G u_h^2 \tag{2.5}$$

For the bubble

$$P_b = P_s + (h_L - s)\rho_L g + \frac{2\sigma}{r} + P_I \tag{2.6}$$

Where P_I is the excess pressure in the bubble due to the liquid inertia as it is moved by the growing and rising bubble. There is some difficulty in determining P_I. Proposed equations are

$$P_I = \rho_L(1.5(\dot{r})^2 + r\ddot{r}) - 0.25\,\rho_L(\dot{s})^2 \quad \text{(McCann \& Prince 1969)}$$

$$P_I = \rho_L(1.5(\dot{r})^2 + r\ddot{r}) \quad \text{(Kupferberg \& Jameson 1969)}$$

$$P_I = 0 \quad \text{(Satyanarayan } et\ al. \text{ 1969)}$$

$$P_I = \rho_L(1.5(\dot{r})^2 + r\ddot{r}) + \frac{\rho_L h_L \ddot{V}_b}{A} \quad \text{(LaNauze \& Harris 1974)}$$

The second term in the last equation accounts for bubble translation (Potter 1969).

As flow rates into and out of the chamber are not necessarily the same, the pressure in the chamber can vary. If adiabatic behaviour is assumed, the pressure in the chamber at time t is

$$P_c - (P_c)_{t=0} = -\frac{c^2 \rho_G}{V_c}\,[V_b - (V_b)_{t=0} - Q_h t] \tag{2.7}$$

Two further equations are

$$u_h = \dot{V}_b / A_h \tag{2.8}$$

and

$$V_b = 4\pi r^3/3 \tag{2.9}$$

Most proposed models have assumed two stages of bubble formation. During the first stage the bubble grows while resting on the plate, either as a complete sphere or as a segment of a sphere. During the second stage the spherical bubble moves away from the plate while still connected to the hole by a small tail.

Generally, it is necessary to assume that $F_R = 0$ and then eqns. (2.1) and (2.5)–(2.9) are sufficient to solve for the six variables P_c, P_b, u_h, r, s, V_b as a function of time over both stages of bubble growth. If the bubble is assumed to be a complete sphere over the first stage, then $s = r$ (Kupferberg & Jameson 1969) and it is unnecessary to assume $F_R = 0$ as the use of eqn. (2.1) is then not required. For segmental sphere first-stage bubbles, the end of the first stage is taken to occur when $s = r$, and for spherical first-stage bubbles it can be taken as the instant when $F_R = 0$ determined by eqn. (2.1). The end of the second stage is usually taken to be when $s = r + r_h$. McCann & Prince (1969) used a slightly different criterion for bubble detachment and also attempted to take into account the wake of the preceding bubble.

2.2.3 *Bubble sizes on formation from single holes*

The bubble volumes on formation measured by a large number of workers have been reviewed by Clift *et al.* (1978) and by D'arcy (1975). The data of McCann & Prince (1969), shown in Fig. 2.7, are typical, and were obtained by measuring the pulsation frequency in the chamber below the plate. Also shown is a comparison with eqn. (2.3) and with predictions using McCann & Prince's own theory which is similar to the general solution outlined above.

As the chamber volume increases, its damping effect increases and the pressure below the hole remains more nearly constant. Consequently, bubbles continue to grow rapidly even up to the point of detachment. As a result a larger chamber volume leads to larger bubble sizes. Eqn. (2.3) represents a lower limit to formation bubble sizes when the chamber volume approaches zero (constant flow regime).

Increasing the pressure increases the gas density and tends to reduce the bubble size on formation as shown by the results of LaNauze & Harris (1974) in Fig. 2.8. A numerical solution of the bubble growth model for this case has been given by Pinczewski (1981).

The formation size bubbles formed from single holes shatter into many small bubbles of varying sizes some 75–100 mm above the hole. Leibson *et*

Fig. 2.7. Bubble volumes on formation from a single hole. Air–water, 1 bar, $d_h = 4.8$ mm. (McCann & Prince 1969.)

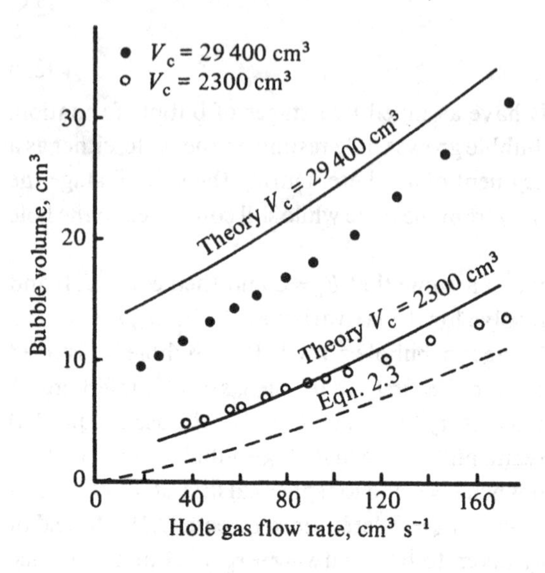

Fig. 2.8. Bubble volumes on formation from a single hole. CO_2–H_2O, $V_c = 375$ cm^3, $d_h = 4.8$ mm. (LaNauze & Harris 1974.)

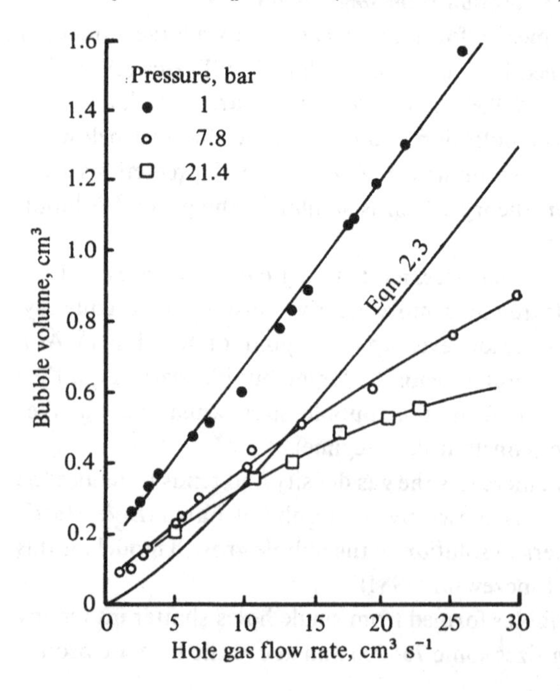

al. (1956) reported that shattering occurred for $Re_h > 2100$. For the air–water system, when $Re_h > 10\,000$, they found that $d_{bs} = 0.0071\, Re_h^{-0.05}$. Similar behaviour was reported by Rennie & Evans (1962) and by Rennie & Smith (1965).

2.3 Bubbling from multiple holes

2.3.1 *Bubble sizes on formation from multiple holes*

For multiple holes, theoretical analysis is complicated by the interaction of the pressure fields around neighbouring holes. However, Kupferberg & Jameson (1970) were able to predict their observed bubble sizes on a sieve tray, having up to 19 holes, using a model similar to the general solution described above for single holes and assuming that each hole bubbled independently of its neighbours. At the low gas flow rates used ($< 9.5\ \mathrm{cm^3\ s^{-1}}$ per hole), not all the holes on the tray bubbled at any instant and they were forced to use experimental values for the fraction of holes bubbling.

As the number of holes increases, the effective chamber volume associated with each hole cannot any longer be considered as isolated from its neighbours. Ultimately, the pressure fluctuations caused by bubbling from any particular hole have little effect on the chamber pressure, and the situation becomes one of constant pressure bubble formation. This has been confirmed experimentally (Haselden & Thorogood 1964, Miyahara *et al.* 1983*b*).

It has often been reported that the bubbling frequency of individual holes on sieve trays is about 20 bubbles $\mathrm{s^{-1}}$. Measurement techniques used have included a resistance probe located in the centre of the hole (Pinczewski & Fell 1972, Hofhuis 1980), measurement of pressure fluctuations near the hole (Porter *et al.* 1967) and photography (West *et al.* 1952, Porter *et al.* 1967). Calderbank (1956) also found the same bubbling frequency from the slots in bubble caps.

Eqn. (2.4) predicts a bubbling frequency of 25 to 16 bubbles $\mathrm{s^{-1}}$ as the gas flow rate per hole varies between 20 and 200 $\mathrm{cm^3\ s^{-1}}$. This agreement can only be regarded as fortuitous. Eqn. (2.4) is derived for constant flow ($V_c = 0$), whereas a sieve tray more nearly corresponds to constant pressure. Furthermore, liquid circulation is neglected in the derivation of eqn. (2.4) and this can have significant influence in determining bubble size on formation.

2.4 Froths

2.4.1 *The bubble regime on sieve trays*

The bubble regime is defined as existing over that range of gas velocities for which the bubbles follow the theory of ideal bubbling. For bubbles rising through a stagnant pool of liquid under ideal bubbling conditions, the relationship between superficial gas velocity u_s and gas holdup ε is (Lockett & Kirkpatrick 1975)

$$u_s = u_{b0}\varepsilon(1-\varepsilon)^{1.39}(1+2.55\,\varepsilon^3) \qquad (2.10)$$

For ideal bubbling, the maximum possible value of u_s occurs at $\varepsilon = 0.6$.

With air bubbles in water, for example, for which $u_{b0} \approx 250$ mm s^{-1}, u_{smax} is 65 mm s^{-1}. In reality factors such as liquid circulation, non-uniformities in bubble size, velocity and spacing, and entrainment of bubbles in each other's wakes, all contribute to a breakdown of ideal bubbling well below $u_s = 65$ mm s^{-1}.

2.4.2 ***The froth regime***

The froth regime is bounded by the bubbling regime at low gas velocities and by the spray regime at high gas velocities. At velocities just above the transition from the bubbling regime, there is a limit to the gas flow rate which can be transported through the liquid by formation size bubbles rising under buoyancy but hindered in their rise by the presence of other bubbles. Liquid circulation can increase this limit to a small extent but, when it is exceeded, the situation is relieved by bubble coalescence which increases the size of some of the bubbles and hence their buoyant rise velocity (u_{b0} in eqn. (2.10)).

Liquid circulation in froths is induced by bubble wakes and particularly by the large bubbles rising through swarms of smaller ones. Liquid circulation also results in a fluctuating turbulent velocity field which influences bubble formation. Depending on the hole spacing and degree of bubble deformation, large-formation-size bubbles can touch and coalesce either during formation or a few millimetres above the tray floor (Miyahara *et al.* 1983*a*, *b*). Furthermore, at industrial flow rates bubble sizes on formation are not uniform and bubble breakup can occur in the intense turbulent velocity field of the froth (Calderbank & Moo-Young 1960).

To summarise, at low gas velocities froths are characterised by a range of bubble sizes and velocities whose characteristics are difficult to predict. As gas velocity increases, gas jetting from the holes plays an increasingly important role. At low froth heights, jets can penetrate completely through the froth and the frequency of this occurrence increases, until at the transition to spray all holes carry jets which penetrate the froth. At any instant the proportion of holes which are bubbling and the proportion having jets penetrating the froth is unknown.

2.4.2.1 *The free bubbling and mixed-froth subregimes*

Hofhuis & Zuiderweg (1979) attempted to identify two subregimes within the froth regime. The free bubbling subregime is defined as existing at relatively low gas velocities where the froth primarily has the characteristics of a bubbling dispersion. The mixed-froth regime exists at higher gas velocities where gas jets and channels are dominant. Ramm (1968) also defined a subregime corresponding to mixed froth which he called 'gas jet and splash'. The transition between the two subregimes has been identified by

$$u_s = (1.15\, g h_w)^{0.5} \quad \text{(Ramm)} \tag{2.11}$$

and

$$u_s = 1.2(g h_{cl})^{0.5} \quad \text{(Hofhuis)} \tag{2.12}$$

The latter equation is recommended. (Note h_{cl} is defined in Section 2.6.) Hofhuis & Zuiderweg (1979) found that Marangoni-induced foam was stable only in the free bubbling regime and that different froth density correlations were necessary in the free bubbling and mixed-froth regimes.

2.4.2.2 *Characteristics of froths*

A description of froth structure requires information on the following:

- diameter and height of the gas jets issuing from the holes which break down to form bubbles;
- bubble size distribution;
- bubble rise velocity distribution;
- proportion of gas passing completely through the froth as jets or channels;
- size of the jets passing through the froth.

Information is available on the first four of the above and is summarised below.

Bubble formation region

Particularly when hole diameter and gas velocity are small and clear liquid height is large, pulsating gas jets issue from the holes and break down to form bubbles before they can penetrate through the froth. Based on very limited data (Lockett, Kirkpatrick & Uddin 1979), the height h_j and diameter d_j of the jets in the bubble formation region can be estimated from

$$h_j \approx 2.9 \times 10^{-6}\, Re_h \quad (Re_h < 10^4) \quad \text{and} \quad d_j \approx 3.4\, d_h$$

The presence of gas jets issuing from the holes in the tray can be clearly seen in Fig. 2.9.

Bubble sizes in froths

Various techniques have been used to estimate bubble sizes in froths.

Photography. Photographs such as Fig. 2.9 show the presence of large bubbles or gas voids (typical vertical dimension 50 mm) rising through a swarm of smaller formation size bubbles. Fig. 2.9 was obtained using a narrow (13 mm) 'two-dimensional' column. Photographs taken through the wall of a normal 'thick' column typically reveal a bubble size at the wall of about 5 mm (West *et al.* 1952, Calderbank & Moo-Young 1960, Calderbank & Rennie 1962, Rennie & Evans 1962, Radionov & Radikovskii 1967, Porter *et al.* 1967, Hobler & Pawelczyk 1972). For many years it was believed that this was the characteristic size of bubbles in froths. Porter *et al.* (1967) convincingly demonstrated that the wall bubbles are unrepresentative. In small-diameter columns, upward liquid flow is induced by large bubbles rising in the centre of the column with a corresponding liquid downflow at the wall. The latter brings small bubbles preferentially to the wall.

Ashley & Haselden (1972) photographed the top surface of a froth from above and observed two distinct ranges of bubble size, $40 < d_b < 80$ mm and $5 < d_b < 10$ mm. They stressed that there was not a continuous distribution of bubble sizes. Similar observations were reported by Hobler & Pawelczyk (1972).

Bubble probe. The bubble probe, shown in Fig. 2.10, was originally developed by Calderbank (Burgess & Calderbank 1975, Calderbank & Pereira 1977, 1979) and was subsequently improved by Raper *et al.* (1977*a*, 1978, 1982). It has been used to determine bubble sizes and velocity distributions in froths. Fig. 2.11 shows the range of bubble sizes reported.

Fig. 2.9. Photograph of froth.

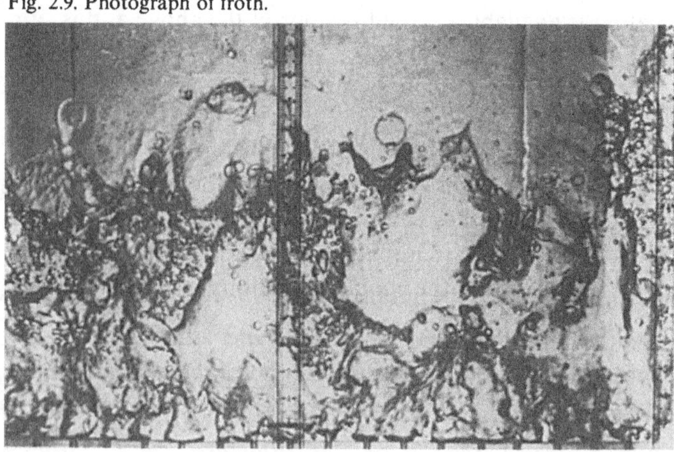

Mass-transfer models. By comparing measured rates of mass transfer with predictions from mass-transfer models in which a minimum value of the mass-transfer coefficient is estimated based on molecular diffusion, it is possible to deduce a minimum possible value for the bubble size. In an outstanding paper, West *et al.* (1952) deduced a minimum bubble diameter of about 13 mm in this way. Similarly, Garner & Porter (1960) argued that the mean bubble diameter in froths must be at least 10 mm to correspond to

Fig. 2.10. The Calderbank bubble probe.

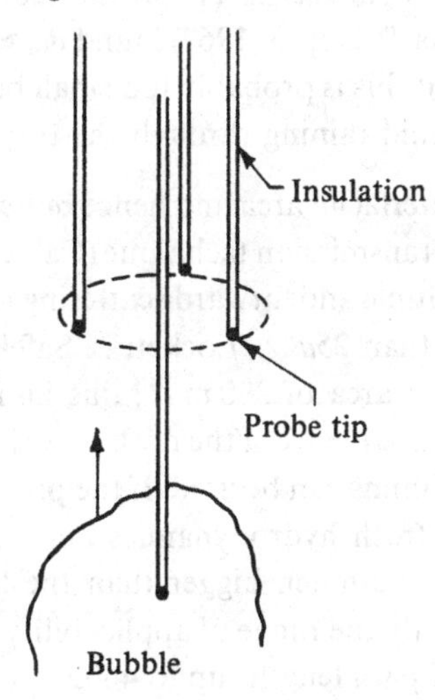

Fig. 2.11. Range of measured Sauter mean bubble diameters in froths vs. hole F factor. (Raper *et al.* 1982.)

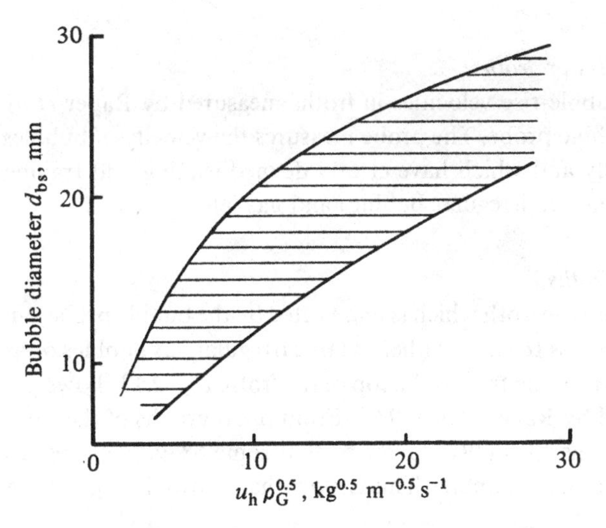

typically observed point efficiencies. Calderbank & Rennie (1962) used a similar argument to infer the presence of gas jetting through the froth.

Interfacial area by mass-transfer techniques. Interfacial area, a, can be measured using mass transfer with chemical reaction and, when coupled with froth density measurements, Sauter mean bubble diameter can be determined from

$$d_{bs} = 6\varepsilon/a \tag{2.13}$$

Some of the difficulties in measuring interfacial area are discussed in Section 8.5.6. Using this technique, Porter *et al.* (1966) quoted d_{bs} as 20–30 mm for bubble-cap trays. Sharma & Gupta (1967) found $d_{bs} \approx 4$ mm for sieve trays without downcomers, but this is probably too small because they neglected mass transfer to the liquid raining through the tray.

Interfacial area by light transmission. Interfacial area and hence bubble size can also be determined using the light-transmission technique (Calderbank 1959). The difficulty is that to avoid multiple and forward scattering of light the optical path usually has to be less than $25a^{-1}$ (Lockett & Safekourdi 1977). For a typical value of interfacial area of 300 m^{-1}, this limits the optical path to less than 83 mm. As a consequence, if the probes are located outside the column, only very small columns can be used. If the probes are located internally, they interfere with froth hydrodynamics and in both cases the optical path length may not be much bigger than the largest bubbles present in the froth. Very recently the range of applicability of the technique has been extended to optical path lengths up to $460a^{-1}$ using a laser source, but so far it has not been used to study froths (Al Taweel *et al.* 1984).

Bubble-rise velocities in froths

Fig. 2.12 shows bubble-rise velocities in froths measured by Raper *et al.* (1977*a*) using a bubble probe. The probe measures the velocity of bubbles which rise vertically and which have clearly defined leading and trailing edges. As such, it misses irregular bubbles and gas jets.

Gas bypassing in froths

Gas passing through the froth which is undetected by the bubble probe can be considered to bypass the froth either as large irregular gas bubbles or as gas jets extending from the tray to the top of the froth. Fig. 2.13 shows gas bypassing reported by Raper *et al.* (1982). From photographs of the froth surface, Ashley & Haselden (1972) deduced that some 68–95% of the gas bypassed the froth as large bubbles. This is more than shown in Fig. 2.13. A possible explanation for the discrepancy is that some of the large bubbles in Ashley & Haselden's study are included in those counted by a bubble probe.

Gas bypassing plays a dominant role in determining point efficiency as discussed in Section 8.3.

2.5 Emulsion flow regime

The discussion of froths so far considered has been primarily concerned with bubble behaviour in stagnant liquids. But on most

Fig. 2.12. Measured bubble rise velocities vs. bubble diameter. $d_h = 6.4$ mm, $\phi = 0.05$, $h_w = 25$ mm. (Raper *et al.* 1977*a*.)

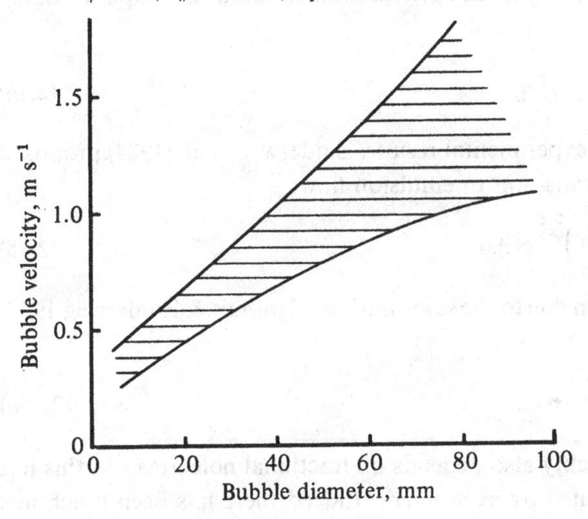

Fig. 2.13. Calculated gas bypassing vs. superficial F factor for sieve, valve and bubble cap trays. (Raper *et al.* 1982.)

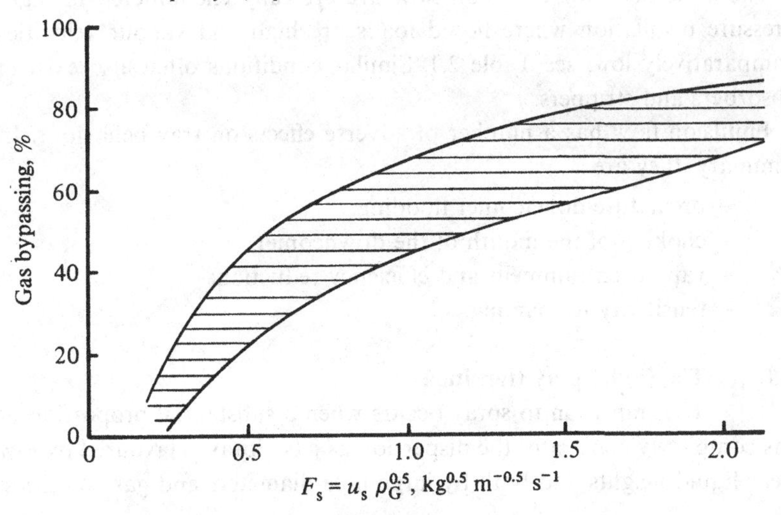

distillation trays the liquid moves horizontally across the tray and velocities (Q_L/Wh_{cl}) as high as 0.5 m s^{-1} are indicated from Table 2.1. As might be expected, liquid flow decreases the size of the bubbles as they are formed because they are sheared off before they can grow to their normal size (Jackson 1952, Kumar & Kuloor 1970). When gas jets are formed at higher gas rates, they can be bent over and broken up into small bubbles by horizontal liquid flow (Zuiderweg *et al.* 1984). The presence of many small bubbles gives the mixture something of the characteristics of an emulsion.

The emulsion flow regime is favoured by a high horizontal liquid momentum compared with the vertical momentum of the vapour issuing from the holes, i.e.

$$\left(\frac{Q_L}{Wh_{cl}}\right)^2 \rho_L > u_h^2 \rho_G \tag{2.14}$$

Based on their own experimental results, Zuiderweg *et al.* (1984) proposed eqn. (2.15) for the transition to emulsion flow:

$$\frac{Q_L}{W}\frac{1}{h_{cl}}\frac{1}{u_s}\left(\frac{\rho_L}{\rho_G}\right)^{0.5} > 3.0 \tag{2.15}$$

An earlier correlation due to the same authors (Hofhuis & Zuiderweg 1979) is

$$\frac{Q_L}{W}\frac{1}{u_s}\left(\frac{\rho_L}{\rho_G}\right)^{0.5} > 0.2 \tag{2.15a}$$

The transition probably also depends on fractional hole area but this has not yet been confirmed by experiment. Indeed there has been much less study of the emulsion flow regime compared with other regimes.

The conditions for emulsion flow are typically encountered in high-pressure distillation where liquid loads are high and vapour velocities comparatively low; see Table 2.1. Similar conditions often also exist in absorbers and strippers.

Emulsion flow has a number of adverse effects on tray behaviour. In summary, they are

- premature downcomer flooding;
- choking of the mouth of the downcomer;
- vapour entrainment and efficiency reduction;
- sensitivity to foaming.

2.6 The froth–spray transition

The transition to spray occurs when a substantial proportion of gas completely penetrates the dispersion as jets. Spray is favoured by low clear liquid heights and both by large hole diameters and gas velocities.

Clear liquid height, h_{cl}, is the volume of liquid in the dispersion per unit bubbling area. It is the liquid depth to which the dispersion would collapse if the gas and liquid flows were instantaneously cut off and weeping of liquid could be prevented.

The proportion of holes carrying gas jets which completely penetrate the froth at the froth–spray transition is open to dispute. Sawistowski has suggested 50% orifice jetting whereas Fell and co-workers (Raper *et al.* 1979) believe 95%+ to be more appropriate. The issue is largely academic since the proportion of holes jetting is difficult to measure and the transition has to be characterised by a change in some other property. However, the hole gas velocity at transition measured by Fell does tend to be higher than other workers consistent with the above.

2.6.1 *Experimental determination of the froth–spray transition*

Measurement techniques which have been used are:

(*a*) *Light beam attenuation.* A horizontal light beam traversing the column is attenuated by spray because the drops intercept the beam. The transition from spray to froth corresponds to an increase in transmitted light providing that the beam is located above the froth surface.

(*b*) *Entrainment rate.* A clearly related technique is measurement of liquid entrainment. A rapid increase in entrainment corresponds to the transition from froth to spray.

(*c*) *Hole pulsation frequency.* An electrical resistance probe located at the centre of a typical hole registers a pulsation frequency of about 20 pulses s^{-1} in the froth regime. The frequency drops to about 5 pulses s^{-1} at the transition to spray which corresponds to a change from bubbling to pulsating jetting at the hole.

(*d*) *Residual pressure drop.* The residual pressure drop h_R is equal to the total tray pressure drop minus clear liquid height minus dry tray pressure drop (see also Section 4.3). For a fixed hole gas velocity, h_R is found to pass through a maximum as clear liquid height is increased. It has been argued (Muller & Prince 1972, Payne & Prince 1975) that h_R reflects momentum transfer from gas to liquid and that it reaches a maximum at the froth–spray transition. The validity of this is by no means clear. Although the residual pressure drop technique is easy to use, what it actually measures is uncertain. Fortunately, it gives results for sieve trays which are in reasonable agreement with those obtained by other techniques. It appears to be suitable only for low free area trays ($<4\%$) (Payne & Prince 1977).

(*e*) *Miscellaneous techniques.* Other techniques which have been used include visual observation, recording the change in the shape of the dispersion density profile (Fig. 2.2), and noting the change in the sound made by the gas.

2.6.2 *Bubbling and jetting at a single hole*

Studies of bubbling and jetting at single holes have helped in the understanding of the mechanism of the froth–spray transition on sieve trays. A regime map for a single 6.4 mm hole using air–water is shown in Fig. 2.14. Muller & Prince (1972) identified line *A* with the froth–spray transition on sieve trays which corresponds to the deformed bubbling–pulsating jet transition. Pinczewski *et al.* (1973) disputed this and proposed an additional region of chain bubbling. They suggested that the appropriate transition was between chain bubbling and pulsating jets, line *B*. Using line *B*, liquid depths from single hole studies correspond to clear liquid heights at transition on sieve trays, whereas if line *A* is used the liquid depths are about three times larger than on sieve trays. Payne & Prince (1975, 1977) overcame the problem by using line *A* and identifying the liquid depth and clear liquid density for a single hole with the froth height and froth density, respectively, on a sieve tray. In this way they were able to bring together transition data from single holes and from sieve trays using a single correlation; see eqn. (2.23). Nevertheless, whether deformed bubbling

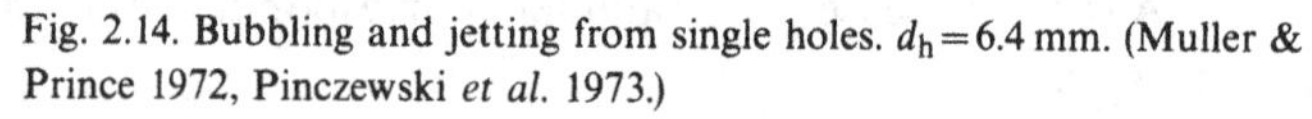
Fig. 2.14. Bubbling and jetting from single holes. $d_h = 6.4$ mm. (Muller & Prince 1972, Pinczewski *et al.* 1973.)

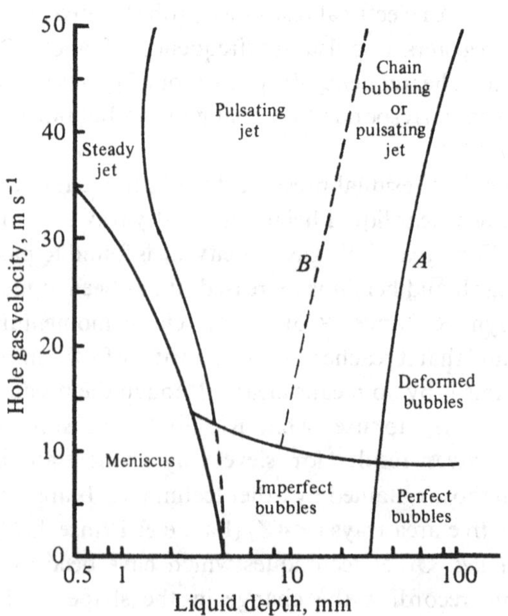

or chain bubbling at a single hole is held to correspond to the froth regime on sieve trays, it is now widely agreed that the froth–spray transition corresponds to a change from bubbling to jetting at individual holes.

2.6.2.1 *Jet penetration model for the bubbling–jetting transition*

The following model is a simple representation of the mechanism governing the froth–spray transition (Lockett 1981). Fig. 2.15*a* shows an idealised gas jet which completely penetrates the froth. The static pressure in the jet is taken to be uniform with height, which may be an oversimplification. The force tending to collapse the jet is the weight of the liquid above it, which tends to cause a liquid bridge and to close the jet, Fig. 2.15*b*. The froth–spray transition occurs when a liquid bridge across the jet is just stable. At the transition, a balance between the force due to gas momentum on the underside of the liquid bridge and the force on the upper side of the bridge due to the weight of liquid above it yields

$$\rho_G u^2 = (h_f - h)\rho_F g \tag{2.16}$$

where u is the gas velocity in the jet at height h and ρ_F is the froth density between h and the froth surface h_f. The gas velocity in the jet falls on moving up from the tray floor because of momentum transfer to the surrounding liquid, and the variation of the diameter of the gas jet with distance from the tray floor can be represented by

$$\left(\frac{d_j}{d_h}\right)^n = a\left(\frac{h}{d_h}\right) + 1 \tag{2.17}$$

Possible variations in pressure due to gas momentum within the jet and pressure within the froth are shown schematically in Fig. 2.16. Uniform froth density is assumed to give the pressure in the froth; see line 1. Curve 2 corresponds to a high-velocity gas stream and to jetting. Curve 4 corresponds to a low-velocity gas stream and to bubbling and curve 3 to the transition velocity. Since line 1 is a tangent to curve 3

Fig. 2.15. (*a*) Gas jet penetrating the froth; (*b*) bridging of gas jet by liquid.

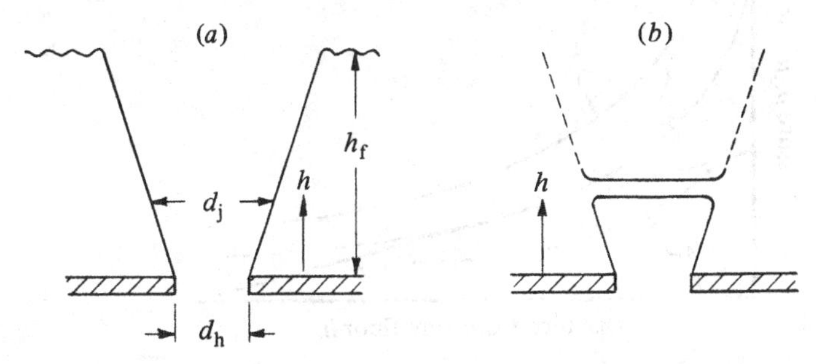

$$\frac{d(\rho_G u^2)}{dh} = -\rho_F g \tag{2.18}$$

For an incompressible gas

$$u d_j^2 = u_h d_h^2 \tag{2.19}$$

Also, $\rho_F = \rho_L(1-\varepsilon)$ and $h_{cl} = h_f(1-\varepsilon)$, where ε is the volume fraction of gas in the froth. From eqns. (2.16)–(2.19),

$$\frac{h_{cl}}{d_h} = \frac{(4+n)/n}{(4/n)^{4/(4+n)}} \left(\frac{\rho_G}{\rho_L}\right)^{n/(4+n)} \frac{u_h^{2n/(4+n)}}{a^{4/(4+n)}} \frac{(1-\varepsilon)^{4/(4+n)}}{(gd_h)^{n/(4+n)}} - \frac{(1-\varepsilon)}{a} \tag{2.20}$$

Eqn. (2.20) gives the clear liquid height at transition, h_{cl}, in terms of the hole gas velocity at transition, u_h. The problem lies in determining a and n. Published froth–spray transition correlations can be classified as corresponding to $n=1$, 2 or 4 in eqn. (2.20).

2.6.3 *Correlations for the froth–spray transition on sieve trays*

2.6.3.1 *Correlations for the froth–spray transition in terms of h_{cl}*

The equation proposed by Barber & Wijn (1979) is

$$\frac{h_{cl}}{d_h} = 1.35\left(\frac{\rho_G}{\rho_L}\right)^{0.25} \frac{u_h^{0.4}}{(gd_h)^{0.2}} \left(\frac{p}{d_h}\right)^{0.33} - 0.59\left(\frac{p}{d_h}\right)^{0.33} \tag{2.21}$$

It was derived assuming a constant cone angle and corresponds to $n=1$ in eqn. (2.20) except for the exponent on (ρ_G/ρ_L). This exponent difference arises because gas inertia was unjustifiably neglected in the derivation of eqn. (2.21).

Fig. 2.16. Possible variations of pressure due to gas momentum in jet and pressure in froth vs. distance from tray surface.

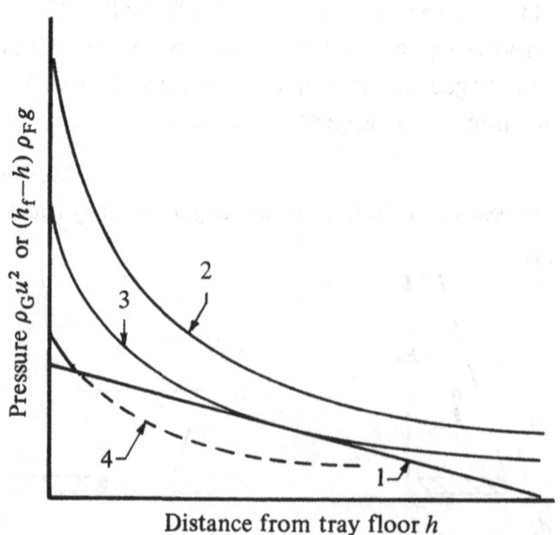

Simple arguments show that the cone angle of the jet must decrease in moving up from the tray, i.e. $n > 1$ (Lockett 1981).

Hofhuis & Zuiderweg (1979) proposed the following purely empirical equation corresponding to $n = 2$ in eqn. (2.20):

$$\frac{h_{cl}}{d_h} = 1.07\left(\frac{\rho_G}{\rho_L}\right)^{0.33} \frac{u_h^{0.66}}{(gd_h)^{0.33}} \tag{2.22}$$

An equation proposed by Payne & Prince (1977) corresponds to $n = 4$ in eqn. (2.20):

$$\frac{h_{cl}}{d_h} = 1.5\left(\frac{\rho_G}{\rho_L}\right)^{0.5} \frac{u_h(1-\varepsilon)^{0.5}}{(gd_h)^{0.5}} \tag{2.23}$$

The picture of jet breakdown proposed by these authors was similar to that of the jet penetration model, the main difference being that the transition from jetting to bubbling was suggested as occurring when the liquid entrainment capability of the gas jet was lower than the rate at which liquid flowed into the jet from the surrounding liquid. The authors did not put this mechanism on a quantitative basis and eqn. (2.23) was simply derived by analogy with the slug-annular flow transition in vertical two-phase pipe flow.

Another equation which bears some resemblance to eqn. (2.20) with $n = 4$ is that of Wong & Kwan (1979):

$$\frac{h_{cl}}{d_h} = 30.5\left(\frac{\rho_G}{\rho_L}\right)^{0.5} \frac{u_h(1-\varepsilon)}{g^{0.5}} + 2.06(1-\varepsilon) \tag{2.24}$$

This is a modification of an earlier equation due to Porter & Wong (1969) which was applicable only to trays of 5% free area. These authors based their transition model on a picture of spray which envisaged the drops to be suspended on the rising gas stream, much as solid particles are suspended in gas fluidised beds. Photographic evidence has shown, however, that this picture of spray is incorrect (Pinczewski & Fell 1974); see Section 2.7. Another early suggestion (Spells & Bakowski 1950, 1952, Spells 1954), that the transition depended on Rayleigh instability of the gas jet, has also proved unfounded.

Detailed comparison of eqns. (2.21)–(2.24) with the available experimental data shows that none of them is completely satisfactory (Lockett 1981). Consequently, little is lost by resorting to empiricism and simply using eqn. (2.25), which is compared with the experimental data in Fig. 2.17:

$$\frac{h_{cl}}{d_h} = 2.78\, u_h\left(\frac{\rho_G}{\rho_L}\right)^{0.5} \tag{2.25}$$

(Note that in eqn. (2.25) u_h must be expressed in m s^{-1}.) The term $(\rho_G/\rho_L)^{0.5}$ arises from the work of Porter & Wong (1969). They, and subsequent workers, also found an insignificant influence of surface tension or liquid viscosity on the transition. Bearing in mind the scatter caused in part by the different techniques used to identify the transition, eqn. (2.25) is reasonably satisfactory.

2.6.3.2 *Correlations for the froth–spray transition avoiding h_{cl}*

In principle any of the correlations of Section 2.6.3.1 can be combined with any of the correlations for h_{cl} in Section 3.4 to predict the transition in terms of the gas and liquid flow rates. The problem is the uncertainty in predicting h_{cl} at the transition point. Loon *et al.* (1973) took a more direct approach as shown in Fig. 2.18, where the transition is

Fig. 2.17. Correlation of froth–spray transition data (Lockett 1981). × Porter & Wong (1969), air–water; ○ Wong (1967), non air–water; ● Pinczewski & Fell (1972); ∇ Payne & Prince (1977), air–water; + Prince *et al.* (1979), distillation; □ Fane & Sawistowski (1969), distillation.

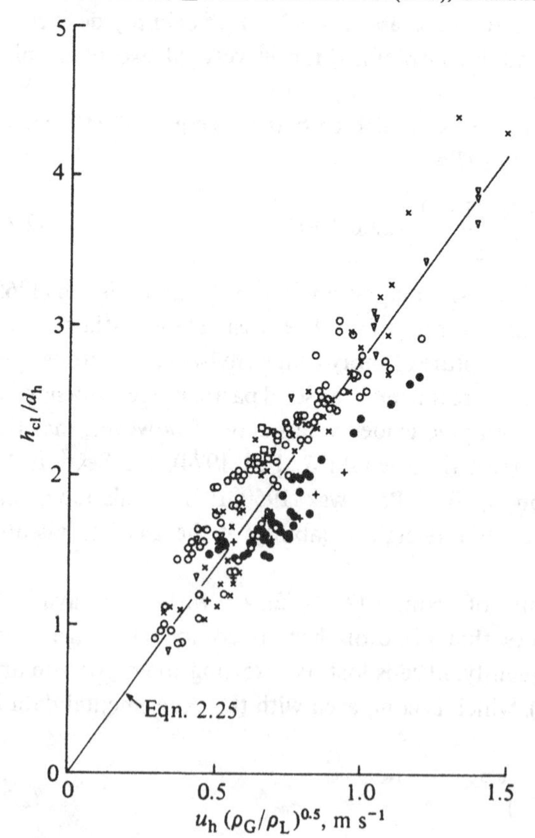

correlated directly as a function of liquid load. Fig. 2.18 can also be represented by the dimensional empirical correlation given by eqn. (2.26):

$$u_s \rho_G^{0.5} = 2.75 \left(\frac{Q_L}{W} \rho_L^{0.5} \right)^n \tag{2.26}$$

where $n = 0.91\,(d_h/\phi)$ (Fell & Pinczewski 1982).

The results on which Fig. 2.18 and eqn. (2.26) were based were obtained using a 25 mm weir, and it was found experimentally that weir height had an insignificant influence on the F_s factor at transition ($F_s = u_s \rho_G^{0.5}$). Unfortunately, a different conclusion is obtained using the alternative approach to predicting the transition involving h_{cl}, since all the h_{cl} correlations available indicate that h_{cl} depends on weir height. Experimental measurements of h_{cl} show that there is a small but finite effect of weir height on h_{cl} at transition conditions. At the present time the situation is unresolved. The two approaches are in reasonable agreement for weir heights of 25 mm, but for higher weirs the correlation of Loon and co-workers predicts an F_s factor at transition lower than other prediction methods which involve an intermediate calculation of h_{cl}.

2.6.4 *Froth–spray transition on trays other than sieve trays*

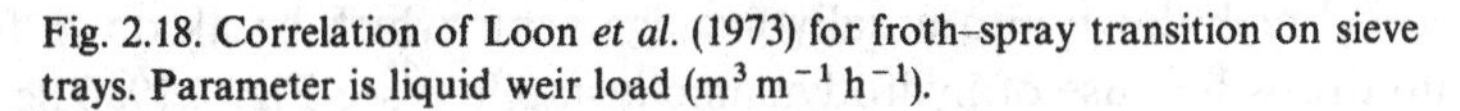

Barber & Wijn (1979) found that eqn. (2.21) was also valid for the froth–spray transition on a proprietary Shell 'snap-in' valve tray. This

Fig. 2.18. Correlation of Loon *et al.* (1973) for froth–spray transition on sieve trays. Parameter is liquid weir load ($m^3\,m^{-1}\,h^{-1}$).

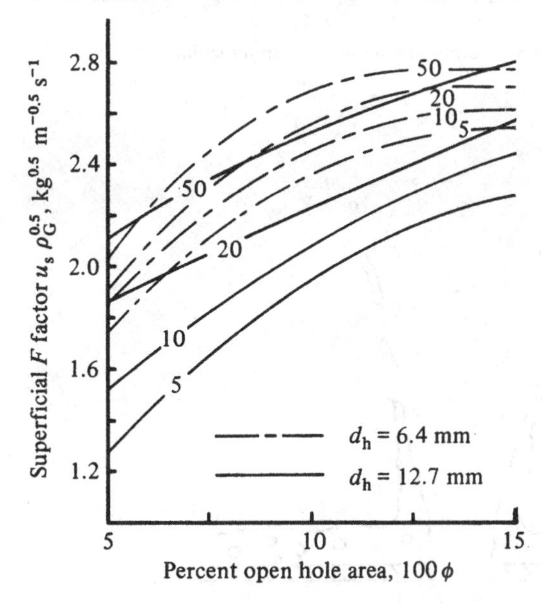

conclusion should be treated with reservations as the transition was determined visually, which is imprecise, and h_{cl} is difficult to measure on valve trays (Section 3.9). Dhulesia (1983) has given transition data for a Glitsch V1 valve tray measured by the light transmission technique.

Spells & Bakowski (Spells & Bakowski 1950, 1952, Spells 1954) demonstrated the existence of two distinct types of bubbling from submerged slots – deep and shallow – which correspond to froth and spray. The existence of a froth–spray transition on bubble cap trays was also reported by Williams *et al.* (1960). However, no attempt appears to have been made to incorporate these observations into a correlation for prediction of the froth–spray transition on bubble cap trays.

2.7 Spray regime

2.7.1 *Introduction*

In spite of some early attempts to describe spray in terms of a fluidisation model (Ho *et al.* 1969, Porter & Wong 1969, Andrew 1969), the free trajectory model has now become accepted (Akselrod & Yusova 1957, Fane & Sawistowski 1969). The current picture is of a liquid, probably containing some small bubbles, which flows over the tray and is penetrated by pulsating gas jets issuing from each hole. Photographic evidence (Nielsen *et al.* 1965, Fane *et al.* 1973, Pinczewski & Fell 1974) shows that the liquid is drawn up into inverted hollow cones by the gas jets. The liquid sheets break down into usually five ligaments which break down further into drops because of hydrodynamic instability; see Fig. 2.19. The drops

Fig. 2.19. Representation of drop formation in the spray regime.

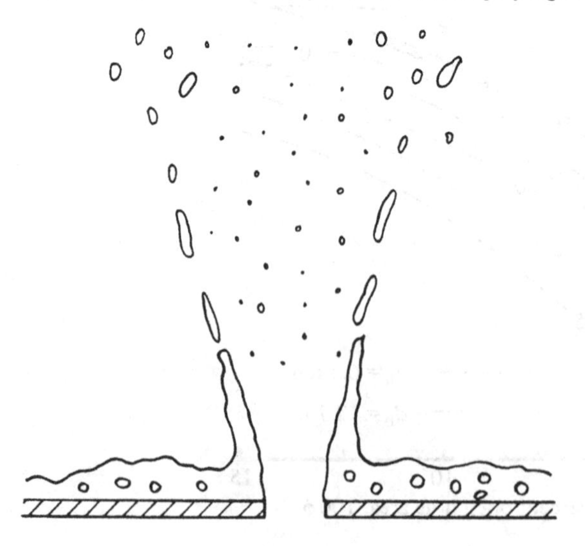

possess an initial projection velocity somewhere between the hole and superficial gas velocities. They are thrown up into the inter-tray space where they describe a free trajectory. The smallest drops may be entrained to the tray above, but the majority fall back to the liquid on the tray to be subsequently re-projected. The spray bed has a very poorly defined upper surface. Photographs have indicated an absence of drop coalescence or breakup during flight (Akselrod & Yusova 1957, Fane *et al.* 1973, Pinczewski & Fell 1977). More recently, indirect and rather inconclusive evidence of drop coalescence has been reported (Raper *et al.* 1979).

2.7.2 *Free trajectory model*

Some success has been achieved in predicting clear liquid height (Raper *et al.* 1979), dispersion density profile (Hai *et al.* 1977, Hofhuis & Zuiderweg 1979) and efficiency (Fane *et al.* 1973) using the free trajectory model. Although unlikely ever to be developed sufficiently for design purposes, the model considerably assists in our understanding of spray.

A vertical force balance on an accelerating drop gives

$$\frac{dv}{dt}=\frac{v\,dv}{dh}=-g\left(\frac{\rho_L-\rho_G}{\rho_L}\right)-\frac{3C_0(v-u_s)|(v-u_s)|\rho_G}{4\rho_L d_p} \tag{2.27}$$

(upwards direction is positive). The gas velocity is assumed to be uniform and given by the superficial velocity. For an initial projection velocity v_p, eqn. (2.27) can be solved to give the drop velocity v as a function of height h.

For a height increment $\Delta h(i)$ at height $h(i)$, the residence time of drop j in the increment can be calculated $\theta(ij)$. (The contributions from the upwards and downwards flights are added.) Given the total rate of drop generation per unit tray area, N_p, and the number fraction $F(j)$ of these drops having size $d_p(j)$, the volume of drops of size $d_p(j)$ in the increment is $\theta(ij)F(j)N_p\pi[d_p(j)]^3/6$.

The volume fraction of all drops at level $h(i)$ is:

$$\alpha(i)=\sum_j \theta(ij)F(j)N_p\pi[d_p(j)]^3/6\,\Delta h(i) \tag{2.28}$$

The total liquid holdup per unit tray area, h_{cl}, above the level at which drops are projected is obtained by summing over all levels:

$$h_{cl}=\sum_i \alpha(i)\,\Delta h(i)=\frac{\pi N_p}{6}\sum_i\sum_j \theta(ij)F(j)[d_p(j)]^3 \tag{2.29}$$

Eliminating N_p from eqns. (2.28) and (2.29),

$$\alpha(i)=\frac{h_{cl}}{\Delta h(i)}\cdot\frac{\sum_j \theta(ij)F(j)[d_p(j)]^3}{\sum_i\sum_j \theta(ij)F(j)[d_p(j)]^3} \tag{2.30}$$

The ultimate purpose of a spray-regime model is to predict hydraulics and mass transfer from first principles. This requires predictions of drop holdup and interfacial area at different heights above the tray. Thus, *a priori* estimates of drop projection velocities, projected drop size distributions and total rate of drop generation are required. These have either been measured directly or have been estimated from the experimental total drop holdup and drop holdup distribution (measured by γ-ray absorption) by parameter fitting using eqns. (2.29) and (2.30).

Jeronimo & Sawistowski (1979) have given a more elegant analytical treatment of the free trajectory model in which they also considered the distribution of interfacial area provided by the drops. By measuring the latter, using light transmission, they claimed that characterisation of the drop projection characteristics was simplified. Only partial success was achieved because it was only possible to determine interfacial area in the upper part of the spray due to the limitation on the light transmission technique that the optical path length should be less than $25a^{-1}$.

2.7.3 *Correlations for use in the free trajectory model*

2.7.3.1 *Diameter of projected drops*

Pinczewski & Fell (1977) and Raper *et al.* (1979) measured drop diameters using both a photographic technique and an electronic probe impact technique. A typical projected drop size distribution for air–water is

Fig. 2.20. Diameter of projected drops in the spray regime. $Q_L/W = 0.0042\ m^2\ s^{-1}$. (Pinczewski & Fell 1977.)

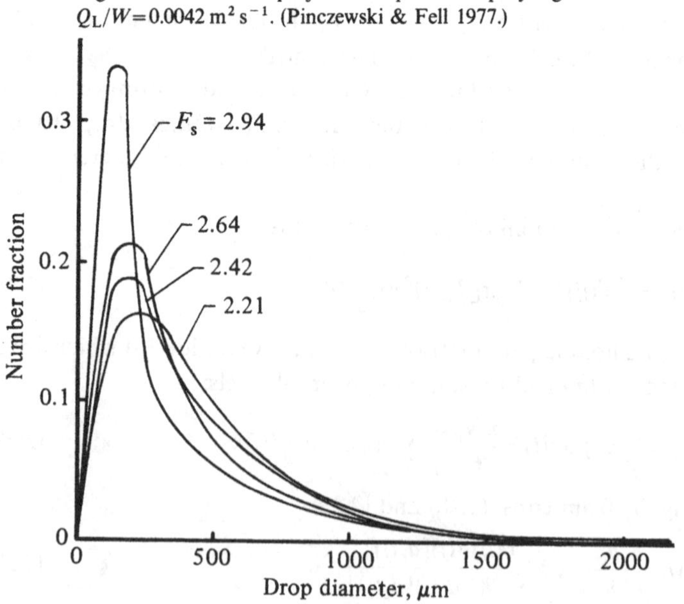

shown in Fig. 2.20. For the Sauter mean projected drop diameter they found $d_{ps} = 0.0315\, u_h^{-0.94}$ and for the projected drop size distribution.

$$\frac{dv}{dy} \doteq \frac{s'}{\sqrt{\pi}} \exp(-s'^2 y^2), \quad \text{with } y = \ln\left(\frac{ad_p}{d_{pm} - d_p}\right) \tag{2.31}$$

Approximately, $d_{pm} = 0.007$ m, $s' = 0.67$ and $a = 0.152\, F_h - 1.517$. Because drops travel at different velocities, there is some difficulty in combining a distribution obtained by sampling over finite time period, as is given by the electronic probe impact technique, with an instantaneous distribution given by photographs. Pinczewski & Fell (1977) overcame this by assuming that the drop velocity was independent of size, but, as shown by Fig. 2.21, this is not strictly correct. In later work (Raper *et al.* 1979) only the electronic probe was used and apparently similar results were obtained. Nevertheless, this does make it difficult to compare probe-determined distributions such as those of Pinczewski with photographically determined distributions such as given by Fane *et al.* (1973).

Fane & Sawistowski (1969) estimated the projected drop size distribution by parameter fitting using the free trajectory model. They recommended the following equations:

$$d_{ps} = 5l(lu_h\rho_L/\mu_L)^{-0.35}(\mu_G u_h/\sigma)^{-0.2} \tag{2.32}$$

and

$$F(j) = \frac{\Delta d_p(j) \exp(-y^2/2s^2)}{\sum_j [\Delta d_p(j) \exp(-y^2/2s^2)]} \tag{2.33}$$

with

$$y = \ln[d_p(j)/d_{ps}]$$

Fig. 2.21. Measured and predicted drop projection velocities in the spray regime. (Raper *et al.* 1979.) ----▲ $d_h = 19.1$ mm; —— ● $d_h = 12.7$ mm; ----× $d_h = 6.4$ mm.

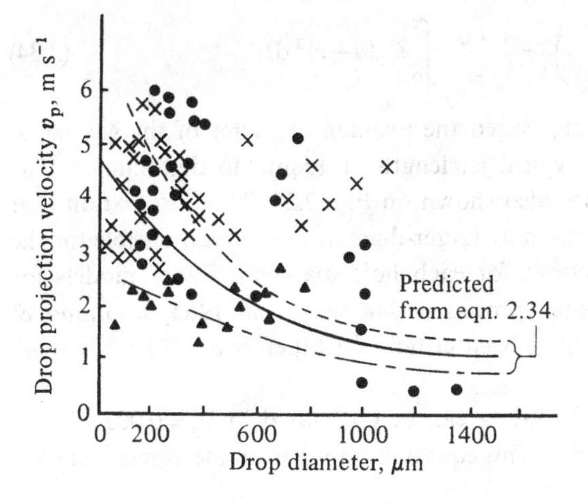

where for surface-tension positive (σ^+) systems (Section 2.8.2.3)

$$l = 2d_h \quad \text{and} \quad s = 0.19$$

and for surface-tension negative (σ^-) systems

$$l = 1.5\, d_h \quad \text{and} \quad s = 0.30$$

These results indicate that σ^+ systems have a larger mean drop diameter than σ^- systems, in agreement with considerations of the Marangoni effect acting in sprays discussed in Section 2.8.2.3.

2.7.3.2 *Drop projection velocities*

Drop projection velocities have been measured photographically (Pinczewski & Fell 1971, Raper *et al.* 1979) and by the plate impingement method (Akselrod & Yusova 1957, Aiba & Yamada 1959). The latter involves measuring the sizes of the spots made on a coated slide held at different heights within the spray. After finding the height at which a particular drop diameter disappears, the corresponding initial drop projection velocity can be determined from the free trajectory model. For a particular drop diameter, there is a distribution of projection velocities and the plate impingement method therefore gives the maximum value. Drop projection velocities determined by Raper and co-workers are shown in Fig. 2.21. Although there is a good deal of scatter, v_p tends to decrease with drop diameter and to increase with hole diameter.

A simple model for prediction of v_p takes the view that it is the velocity acquired by the drop immediately after sheet breakdown in the high-velocity gas jet which issues from the hole (Pinczewski & Fell 1971, Raper *et al.* 1979). Thus, if the drop is accelerated for time t, it follows from eqn. (2.27) that

$$v_p = -g\left(\frac{\rho_L - \rho_G}{\rho_L}\right)t + \frac{3}{4}\frac{\rho_G}{d_p \rho_L}\int_0^t C_0 (u - v)^2\, dt \qquad (2.34)$$

Raper and co-workers measured the average diameter of the gas jet to estimate u and took a typical jet length of 15 mm to determine t. The predicted values of v_p are also shown on Fig. 2.21. The observation that small holes have proportionally larger-diameter jets is responsible for the difference in the predictions for each hole diameter. Other models for prediction of v_p have been proposed (Nielsen *et al.* 1965, Jeronimo & Sawistowski 1974) but have been criticised (Raper *et al.* 1979, Ho *et al.* 1969).

Aiba & Yamada's (1959) data can be correlated by $v_p = 0.0024\, d_p^{-0.93}$ (Fane & Sawistowski 1969). This equation is in reasonable agreement with the results shown in Fig. 2.21. An alternative equation due to Manickampillai & Sawistowski (1981) is

$$v_p = 0.01795\, d_h^{0.75} u_h^2 d_p^{-0.83} \mu_L^{-0.2} \sigma^{-0.3} \rho_G \rho_L^{-1} \tag{2.35}$$

2.7.3.3 *Total rate of drop generation*

Raper *et al.* (1979) estimated N_p from measurements of the entrainment generated at the hole for a single hole, Fig. 2.22, and assumed that it also gave the rate of drop generation per hole on a sieve tray. Much more work is required before N_p can be estimated confidently in this way.

2.7.3.4 *Liquid holdup profile in the spray regime*

The liquid holdup profile passes through a characteristic maximum in the spray regime, as shown in Fig. 2.2, and it can be predicted quite successfully using the free trajectory model – eqn. (2.30) (Hai *et al.* 1977). This achievement is tempered by noting that the experimental value of h_{cl} is required in eqn. (2.30) – in effect forcing the areas under the predicted and experimental curves to be identical. The maximum in the profile results from the distribution of drop diameters and projection velocities.

A neglected difficulty is that the model predicts liquid holdup only in the free trajectory zone, whereas the experimentally determined h_{cl} also includes the drop propagation zone close to the tray. However, Raper *et al.* (1979) were able to predict the liquid holdup profile, without having to measure h_{cl}, using eqn. (2.28) together with projected drop characteristics from, for example, Figs. 2.20, 2.21 and 2.22. Hofhuis & Zuiderweg (1979) also successfully predicted liquid holdup profiles in the mixed-froth regime using a free trajectory model.

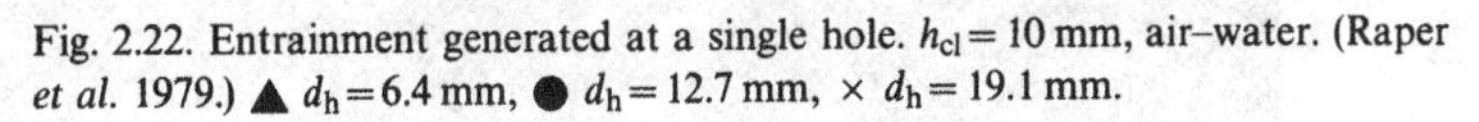

Fig. 2.22. Entrainment generated at a single hole. $h_{cl} = 10$ mm, air–water. (Raper *et al.* 1979.) ▲ $d_h = 6.4$ mm, ● $d_h = 12.7$ mm, × $d_h = 19.1$ mm.

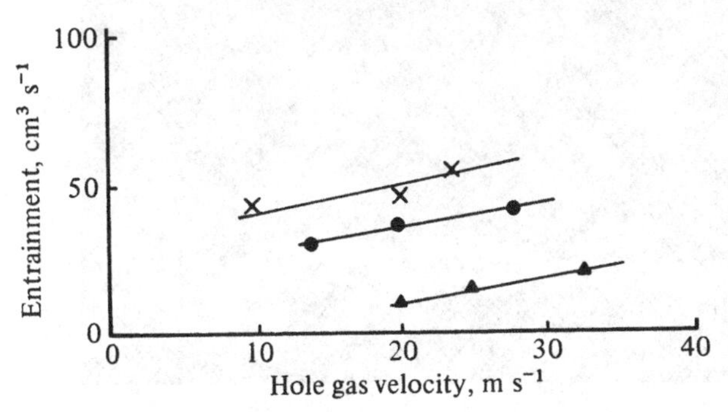

2.8 Foam

2.8.1 *Introduction*

It can be shown thermodynamically that, in the absence of heat or mass transfer, *pure* liquids cannot sustain a stable foam (Ross 1967). When foaming occurs, it follows that it results either from the presence of mixtures, impurities, or heat or mass transfer, which convey on bubbles the ability to resist coalescence either with each other or at the upper surface of the dispersion. A resistance to bubble–bubble coalescence results in a smaller mean bubble size, so that the bubbles rise slower and increase the gas holdup in the dispersion. A resistance to bubble–surface coalescence results in an accumulation of bubbles at the surface which have a finite drainage time before coalescence eventually occurs. The bubble blanket can extend downwards to fill the liquid with draining bubbles. As there is less liquid trapped between slowly draining bubbles than between the bubbles in a froth, the foam height rises compared with that of a froth to accommodate the total liquid in the system.

Bubble coalescence proceeds by drainage of the intervening liquid film. A foam will tend to form when a mechanism exists to maintain the film and to

Fig. 2.23. Unstable foam.

prevent it rupturing prematurely during the drainage process. The stability of the foam depends on the ability of the film to heal itself against excessive localised thinning as overall film drainage proceeds. A range of foam types can be obtained, depending on the degree of film stability, ranging from unstable through to metastable type foam. In unstable foam the bubbles jostle together as spheres (Kitchener & Cooper 1959). If film stability is only slight, unstable foam is barely distinguishable from froth. Metastable foams persist much longer and the bubbles become distorted into pentagonal dodecahedra to give what is usually called cellular foam; see Figs. 2.23 and 2.24.

2.8.2 ***Film stabilisation by the Marangoni effect***

2.8.2.1 *Causes of the Marangoni effect*

The commonest mechanism conferring film stability is the Marangoni effect (Marangoni 1871) which can arise in several ways. If a surface-active solute is present which reduces the surface tension, it tends to concentrate at the liquid surface because by so doing it minimises the free surface energy of the system. If, during film drainage, excessive localised film

Fig. 2.24. Metastable cellular foam.

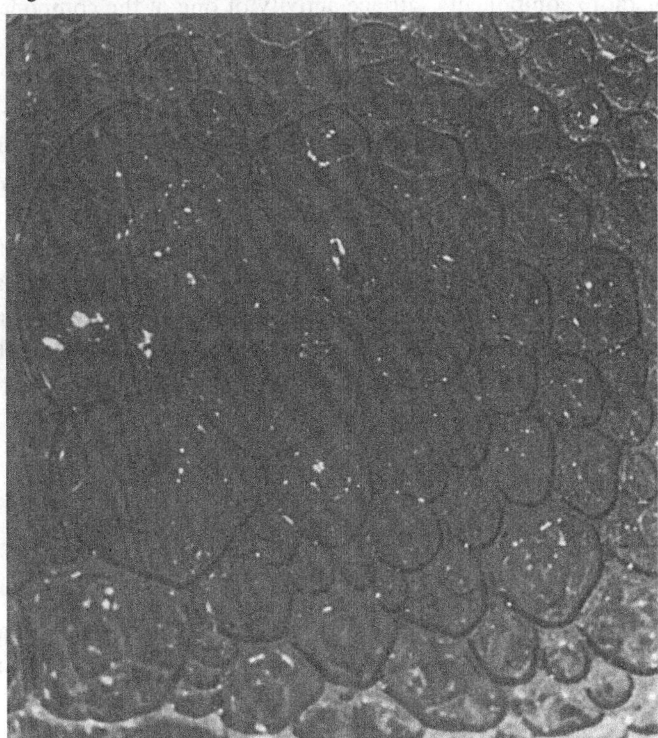

thinning occurs, it results in a local increase of film surface area; see Fig. 2.25. In the thin part of the film (*A*), the surface concentration of the surface-active solute is reduced with a resulting rise in the local surface tension. Alternatively, film drainage itself can cause a local increase in surface tension, because as the liquid in the film flows, it can drag along some of the surface layer, thereby locally reducing the surface concentration of the surface-active solute. Usually diffusion of the surface-active solute from the bulk film to the surface is too slow for it to play any part in replenishing the surface deficiency. Consequently, a net surface force develops acting towards the region of high surface tension. This causes a counter surface flow of liquid (*B* to *A*) which counteracts film drainage and restores the film.

The presence of surface-active solutes can arise from adventitious surface-active impurities. But even in the absence of impurities, Andrew (1960) showed that, when one of the normal constituents of the mixture has a lower surface tension than the others, it will tend to concentrate at the surface and stabilise the film by the Marangoni effect.

2.8.2.2 *Ross-type foams*

Ross & Nishioka (1975) have argued that in a distillation system of two or more components, surface activity of one of the components is a result of a weak interaction with the others. Thus, a tendency towards the separation of two liquid phases serves as an indication that surface activity, and hence foaming, is likely to occur. They suggest that this may explain why foaming is often a problem in extractive distillation columns in which liquid phase non-ideality, while increasing the relative volatility of the components to be separated, also leads to a tendency towards the separation of two liquid phases. They vividly demonstrated that as the mixture composition approaches the critical point (two components) or the plait point (three or more components) foaming can be severe. Once two liquid phases appear, however, one of the phases can act to destroy the foam if the separated phase is present in small quantity and it has a lower surface tension than the other phase.

Fig. 2.25. Maragoni effect in the presence of surface-active solute.

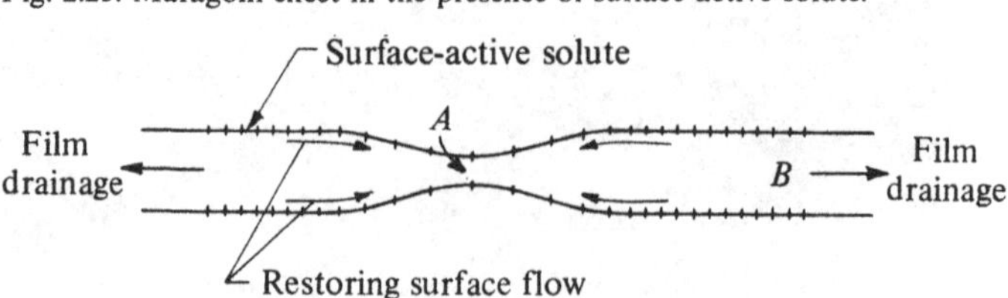

2.8.2.3 *Mass-transfer-induced Marangoni effect*

The most celebrated example of foaming caused by the Marangoni effect arises because of mass transfer. Zuiderweg & Harmens (1958) classified distillation and absorption systems into three types:

(*a*) Positive systems, σ^+, where the surface tension increases as the liquid proceeds down the column.

(*b*) Negative systems, σ^-, where the surface tension falls as the liquid proceeds down the column.

(*c*) Neutral systems, σ^0, where the surface tension is virtually unchanged from tray to tray. This can arise either because the components have similar surface tensions or because the composition change, and hence the surface tension change, from tray to tray is small.

During distillation, the liquid in the thin film between the bubbles (Fig. 2.26) is richer in the less-volatile components compared with the rest of the liquid surrounding the bubbles. During absorption it has a higher concentration of absorbed gas. For a σ^+ system, the liquid in the film is therefore of higher surface tension and a surface-tension gradient results along the surface of the film which causes a surface flow of liquid into the film. This thickens and locally reinforces thin parts of the film, and consequently σ^+ systems have a tendency to foam. Conversely, for σ^- systems, there is a surface flow out of the film which acts to thin and break the film. Thus σ^- and σ^0 systems do not foam to any significant extent.

Sawistowski and co-workers (Bainbridge & Sawistowski 1964, Fane & Sawistowski 1968) pointed out that the Marangoni effect operates during the breakdown of liquid ligaments to form drops in the spray regime. By a

Fig. 2.26. Mass-transfer-induced Marangoni effect.

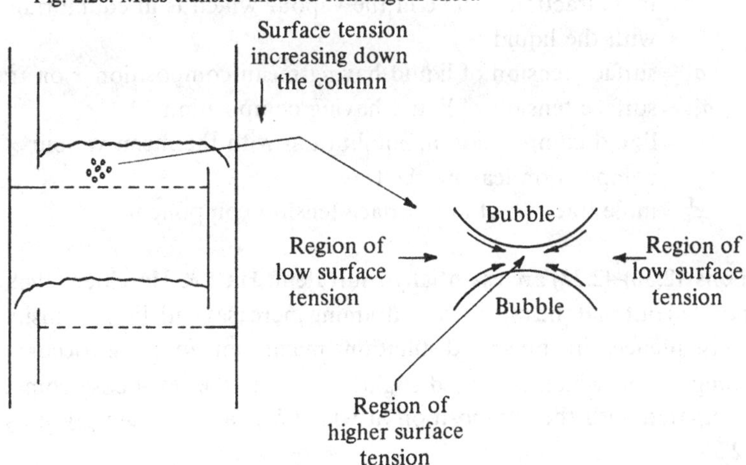

similar mechanism to that described above, ligaments are stabilised in σ^+ systems and ligament breakdown is assisted in σ^- systems. Thus, smaller and more numerous drops are expected in σ^- systems operating in the spray regime in accord with the mean drop size in sprays discussed in Section 2.7.3.

2.8.2.4 *Stabilisation indices*

The importance of the Marangoni effect in determining foaming depends both on the change of surface tension with composition and the composition change involved. Thus maximum foaming has been correlated with a maximum in the following stabilisation indices S:

when mass transfer occurs

$$S=\frac{\partial\sigma}{\partial x}(y^*-y) \quad \text{(Hart \& Haselden 1969)} \tag{2.36}$$

$$S=\sigma_{x^*}-\sigma_x \quad \text{(Lowry \& Van Winkle 1969)} \tag{2.37}$$

$$S=\frac{\partial\sigma}{\partial x}(x-x^*) \quad \text{(Hovestreydt 1963)} \tag{2.38}$$

in the absence of mass transfer

$$S=x^{\mathrm{L}}\left(\frac{\mathrm{d}\sigma}{\mathrm{d}x^{\mathrm{L}}}\right)^2 \quad \text{(Andrew 1960)} \tag{2.39}$$

where

σ = surface tension
x = mole fraction of mvc in liquid (mvc = more volatile component)
y = mole fraction of mvc in vapour
y^* = mole fraction of mvc in the vapour which is in equilibrium with the liquid
σ_x = surface tension of liquid having mean composition x on tray
σ_{x^*} = surface tension of liquid having composition x^*
x^* = liquid composition in equilibrium with the mean vapour composition leaving the tray
x^{L} = mole fraction of low surface tension component

Eqns. (2.36)–(2.38) are essentially equivalent. Hart & Haselden (1969) also pointed out that the rate of foam draining increases with liquid density. As a consequence, in binary distillation maximum foaming occurs at a composition which is shifted slightly towards the less-dense component compared with the composition at which S is a maximum given by eqn. (2.36).

2.8.3 *Other causes of film stabilisation*

Gelatinous surface layers. Ross (1967) attributed the existence of very stable foams (e.g. beer foam) to the existence of gelatinous surface layers in the film. These immobilise the liquid so that the normal pressure differences caused by gravity and capillary action are insufficient to promote flow in the film. The gelatinous surface film exhibits a sharp transition to a freely flowing film over a narrow temperature range so that these foams can break down rapidly as the temperature is raised.

Finely divided solids. Solid particles can stabilise foam films. Bikerman (1973) has shown that there is usually an optimum size and wettability of the particles which gives maximum foam stability.

Bubble contact velocity. Bubble coalescence can be inhibited by a large bubble contact velocity (Kirkpatrick & Lockett 1974). Bubble bounce can occur before the intervening liquid film has time to drain.

2.8.4 *Hydrodynamic conditions for foam stability on trays*

Foaming does not occur on a tray operating in the spray regime. This is due to the limited number of bubbles separated by liquid films and also to the large gas inertia which mechanically disrupts any foam which tends to form. Thus Rennie & Evans (1962) found that foam tends to break down when the hole Reynolds number is greater than 2100. Similarly, Hofhuis & Zuiderweg (1979) defined the mixed-froth regime as a high gas velocity subregime within the froth regime in which Marangoni foam is unstable. De Goederen (1965) found that foam broke down above $u_s = 0.5\ \mathrm{m\ s^{-1}}$ and a number of Russian studies have indicated that foaming is favoured by low gas velocities (Pozin *et al.* 1957).

Hofhuis & Zuiderweg (1979) have also proposed that foaming occurs primarily in the emulsion regime at high liquid-to-gas ratios. However, even if the tray operates in the mixed-froth or spray regime, it should be noted that foam can still be generated in the downcomers, particularly if mass transfer occurs there. Ross-type foam does not require mass transfer to sustain it and so can be expected to persist more readily in the downcomers.

When foaming occurs on industrial trays, the liquid is likely to have a significant horizontal velocity. The relevance of academic studies in which stagnant liquid was used can therefore be questioned. Some early work (Calderbank & Rennie 1962) in which flowing liquid was used can be criticised because the column diameter was small (0.1 m). The liquid was introduced near the tray floor and was pumped upwards by the gas to leave the column over a weir. The arrangement has been criticised (Sargent & McMillan 1962) because it fails to reproduce the situation on large-diameter trays where the liquid which finds its way to the top of the foam has to drain down again against the gas flow.

2.8.5 ***Mathematical models for foam***

2.8.5.1 *Cellular foam*

Models of increasing complexity have been proposed for cellular foam (see, for example, Barber & Hartland 1975 and Steiner *et al.* 1977). The foam bubbles are taken to be regular dodecahedrons with pentagonal sides (Fig. 2.27). Liquid is contained in the pentagonal films and in the Plateau borders which are the channels formed at the sides of the pentagons where the edges of three bubbles meet. Because of their curved walls, the pressure in the Plateau borders is lower than in the films so that liquid drains from the films into the borders. On a distillation tray there is no net vertical liquid flow, so the liquid which is carried up in the foam as the bubbles move upward drains down again under gravity in the Plateau borders (and possibly also in the films (Steiner *et al.* 1977)). Eventually, as they move upward, the films drain sufficiently to rupture, so defining the top of the foam. Although foam height is of primary interest to the tray designer, the models are not sufficiently well developed for it to be predicted with any confidence. However, the models have been used successfully to predict the liquid holdup profile in the foam (Steiner *et al.* 1977). Some unresolved difficulties in using the models are in predicting the initial bubble size and the frequency of bubble coalescence, in allowing for surface viscosity in the film and border drainage equations, and in determining the film thickness at rupture.

The cellular foam–froth transition is complex and a reliable correlation is not available. It is, however, characteristically accompanied by the onset of liquid circulation (Porter *et al.* 1967, Ho *et al.* 1969). The reason for this is that bubbles in a cellular foam are displaced upwards by forming bubbles at

Fig. 2.27. Regular pentagonal dodecahedral bubble in cellular foam.

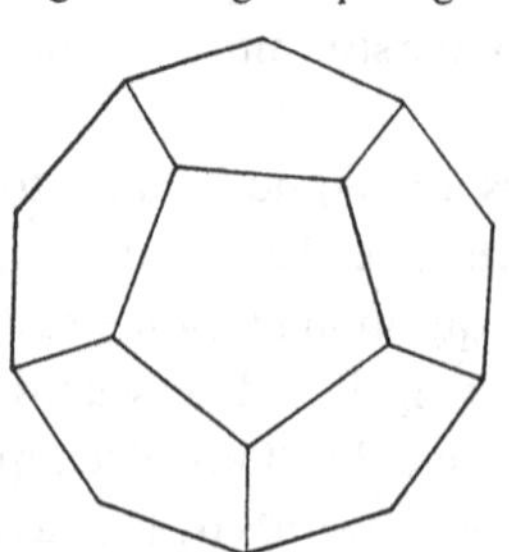

a velocity close to the superficial gas velocity ($u_b = u_s/\varepsilon$ and $\varepsilon > 0.8$). On transition to froth, if the liquid remained stagnant, the bubbles would rise under buoyancy at a much reduced velocity because of bubble interaction (eqn. (2.10)). Liquid circulation starts on transition to froth so as to achieve the necessary gas throughput.

2.8.5.2 *Unstable foam*

In contrast to the above, it has been suggested (Andrew 1960) that unstable foams in which relative bubble movement occurs are more characteristic of the foams found in practice. Barber & Wijn (1979) developed a model for unstable foams and applied it to foaming in flash vessels and in downcomers. The basic idea behind their model is that, as the bubbles are forced together by the turbulent eddies in the mobile foam, the intervening liquid film drains and bubble coalescence occurs. Eventually the bubbles grow large enough to escape from the foam. The total time for the required number of coalescence steps to occur determines the bubble residence time and hence the mean gas holdup in the foam. The model gives reasonable predictions of the foam height in flash vessels but is less successful when used for downcomers. The approach has not yet been used for mobile foams on trays.

2.8.6 ***Dealing with foaming systems***

2.8.6.1 *Foam and tray design*

A minor degree of foaming can raise tray efficiency by increasing the gas–liquid contact time (Zuiderweg & Harmens 1958, Brumbaugh & Berg 1973). However, excessive foaming usually increases entrainment and downcomer backup and eventually leads to premature flooding. Unfortunately, although the factors which lead to film stabilisation, and hence foaming, are reasonably well understood qualitatively, prediction of their quantitative importance in practical situations is difficult. Consequently, past experience is usually used to identify troublesome foaming systems, sometimes supplemented by foam tests in a 'foam cell'. A typical foam test involves bubbling nitrogen or other gas through a liquid sample in a column a few centimetres in diameter. While this can identify foaming due to surface activity or suspended solids, it gives limited information about Marangoni foaming caused by mass transfer. The latter can be identified using an Oldershaw column but, because of the low vapour velocity used in such a column, the extent of foaming may be overestimated. The best, but most expensive, way to test for foaming is to use a pilot column in which the hydrodynamics and physical properties are as close as possible to those in the actual column. Having identified a potential foaming problem, it is usually dealt with by derating the design of the tray and downcomer. Deliberate design to ensure operation in the spray regime can also be used. When these alternatives are not possible, for example when foaming occurs unexpectedly in an existing column, it is usual to use an antifoaming agent.

Derating is achieved by using empirical system factors, some of which are listed in Table 2.2. Their use is dealt with in Chapter 5. Apparently, there has been no published attempt to relate these system factors to the various mechanisms which stabilise the liquid films. Published accounts of foaming in industrial columns, where the mechanism of foaming has been identified, are discussed below.

Marangoni foaming by mass transfer occurs during the distillation of air. Thus O_2–N_2 and Ar–N_2 are both σ^+ systems and foam, whereas O_2–Ar is σ^0 and does not foam (Haselden & Thorogood 1964). The same type of foaming has been reported in the extractive distillation of C_5 hydrocarbons using acetonitrile–water as a solvent, during the stripping of paraffins and

Table 2.2. *System factors*

	System factor *SF*
Slight foaming	
Depropanisers	0.9
Freons	0.9
H_2S strippers	0.9
Hot carbonate strippers	0.9
Moderate foaming	
De-ethanisers	0.85
Oil absorbers	0.85
Amine strippers	0.85
Glycol strippers	0.85
Sulpholane systems	0.85
Crude towers	0.85
Hot carbonate absorbers	0.85
Furfural refining	0.80
Heavy foaming	
Amine absorbers	0.75
Glycol contactors	0.65
Methylethyl ketone	0.60
Stable foam	
Alcohol synthesis absorbers	0.35
Caustic regenerators	0.3

naphthenes from aromatics using sulpholane (van der Meer 1971, van der Meer *et al.* 1971), and during alcohol–water separation (Zuiderweg & Harmens 1958, Zuiderweg 1983). In the ethanol–water system protein impurities may also lead to foam (Danckwerts *et al.* 1960).

Bolles (1967) described foaming during the separation of olefins in a reboiled absorber using DMF as a solvent. This has been variously attributed to Marangoni foaming by mass transfer (van der Meer 1971) and to the formation of gelatinous surface layers (Ross 1967).

Foaming occurs in the columns of the GS process for producing heavy water. One factor which may contribute to this is stabilisation of the liquid films by adsorbed H_2S (Sagert & Quinn 1976).

Barber & Wijn (1979) have attributed foaming in the flash vessel and on the trays of a crude distillation column to surface-active components and to the presence of fine solid particles.

2.8.6.2 *Antifoaming agents*

Some disadvantages of antifoaming agents are their cost and the possibility of product contamination. Theoretical explanations of how they work have been given by Ross (1967), Bikerman (1973) and others. They can be divided into two types: foam preventatives, which prevent foam forming, and foam breakers, which are added to existing foams.

Foam preventatives are believed to adsorb at the film surface in preference to the film stabilising surface-active constituents. They can reduce the 'coherence' of the film surface, can increase the solubility of the film-stabilising constituents in the bulk liquid, or, if they diffuse very rapidly from the bulk to the surface, can annul any transient rise in local surface tension before the Marangoni effect has time to act. In addition it has been suggested (van der Meer 1971) that foam preventatives can cause a large reduction in the overall surface tension such that surface tension gradients along the film are practically eliminated.

A foam breaker acts by spreading over the surface of the liquid film, carrying with it a layer of the underlying liquid, so thinning and breaking the film.

Theoretical understanding of foam inhibition and breaking is by no means complete, and consequently selection of an antifoaming agent can be a hit or miss process. Silicones (polydimethylsiloxanes) are common antifoaming agents in tray columns. An exhaustive compilation of industrial antifoaming agents is available (Kerner 1976). An interesting technique for injecting antifoaming agents as an aerosol spray has been described by Pratt & Hobbs (1975).

3

Clear liquid height, dispersion height and density

3.1 Introduction

This chapter is concerned with prediction of the height of the froth or spray bed, with its average density and with prediction of the clear liquid height. The latter occurs in nearly every hydraulics correlation and is perhaps, after hole gas velocity, the single most important variable influencing hydraulic behaviour. Fig. 2.2 shows that the dispersion has no distinct upper surface and this is particularly true for spray. So visual determination of dispersion height is highly inaccurate. Using γ-ray absorption, the dispersion height can be defined as the height where the liquid holdup fraction falls to an arbitrary value. For spray 0.01 has been used (Hofhuis & Zuiderweg 1979) and 0.1 has been used for froth (Lockett *et al.* 1979).

The average liquid holdup fraction α is related to the dispersion height h_f by

$$\alpha = h_{cl}/h_f \tag{3.1}$$

The average gas holdup fraction ε is

$$\varepsilon = 1 - \alpha \tag{3.2}$$

The average dispersion density ρ_F is given by

$$\rho_F = \alpha\rho_L + (1-\alpha)\rho_G \tag{3.3}$$

3.2 Measurement of clear liquid height

Clear liquid height can be accurately determined on sieve trays by integration of vertical liquid holdup profiles such as those of Fig. 2.2. This is time consuming and tedious and other methods are usually preferred. An alternative is to subtract the dry tray pressure drop from the total tray pressure drop – eqn. (4.7). This approach is commonly used for industrial trays but it lacks precision because it neglects the residual pressure drop –

see Section 4.3. Consequently, the usual method used is to determine h_{cl} from the measured pressure at the tray floor using a manometer. Fig. 3.1 shows the arrangement used. Clear liquid height is related to the manometer reading through a vertical momentum balance between sections 1 and 2:

$$nu_h^2 A_h \rho_G + P_1 A_b = u_s^2 A_b \rho_G + P_2 A_b + h_f A_b g(\alpha \rho_L + (1-\alpha)\rho_G) + F_w \quad (3.4)$$

In eqn. (3.4) n = number of holes in area A_b, P_1 = pressure at tray floor assumed uniform over the tray, P_2 = pressure above the dispersion where the gas velocity is u_s, and F_w = the frictional force due to the walls.

If $\rho_G \ll \rho_L$, neglecting F_w, and since $nA_h u_h = A_b u_s$ from eqns. (3.1) and (3.4),

$$P_1 - P_2 = h_{cl}\rho_L g - u_s \rho_G (u_h - u_s) \quad (3.5)$$

The dynamic liquid head is defined as

$$h_{clD} = \frac{(P_1 - P_2)}{\rho_L g} \quad (3.6)$$

so that

$$h_{clD} = h_{cl} - \frac{u_s \rho_G (u_h - u_s)}{\rho_L g} \quad (3.7)$$

The clear liquid height is related to the manometer reading h_m by correcting for capillary rise, so that

$$h_{cl} = h_m + \frac{u_s \rho_G (u_h - u_s)}{\rho_L g} - \frac{2\sigma}{\rho_L g r_m} \quad (3.8)$$

Many reported values of h_{cl} in the literature are in fact values of h_{clD} because the gas momentum correction of eqn. (3.8) was neglected.

Eqn. (3.8) has been verified by comparing h_{cl} measured by manometer against integrated froth density profiles obtained by γ-ray absorption (Prince *et al.* 1979) and against h_{cl} fixed by using a measured amount of

Fig. 3.1. Measurement of clear liquid height by manometer.

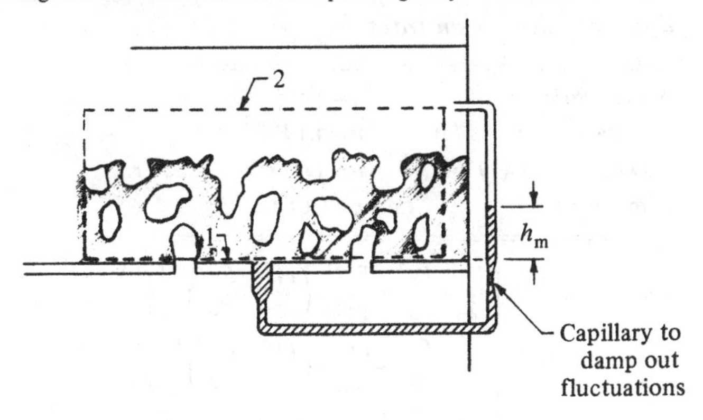

liquid on a tray with no liquid cross flow (Davies & Porter 1965). Some care is required in small columns because of liquid circulation (up in the centre, down at the sides) and h_{cl} is generally smaller at the centre of such columns than the average over the tray (Davies & Porter 1965). A recent test of eqn. (3.8) using a single centrally located manometer in a small column (Bennett *et al.* 1983) must be considered inconclusive because of the probability of liquid circulation.

3.3 Prediction of dispersion density

Theoretical investigations of the conditions under which the energy of the two-phase dispersion on the tray is at a minimum have led to predictions of the vertical liquid holdup profiles (Azbel 1963, Kim 1966, Kolar 1969, Livansky & Kolar 1971, Cervenka & Kolar 1973*a*, Takahashi *et al.* 1973, Steiner *et al.* 1975, Kawagoe *et al.* 1976, Unno & Inoue 1976, Takahashi *et al.* 1979). Unfortunately, these bear little resemblance to the complex profiles in the froth and spray regimes determined experimentally and shown in Fig. 2.2. However, a useful outcome of these theoretical studies, particularly those based on the work of Azbel, is that they predict that the mean liquid holdup fraction α is a function solely of the Froude number Fr, Table 3.1, where

$$Fr = \frac{u_s^2}{g h_{cl}} \tag{3.9}$$

and

$$\eta = \frac{\varepsilon}{1-\varepsilon} = \frac{1-\alpha}{\alpha} \tag{3.10}$$

Not surprisingly, the purely theoretical equations shown in Table 3.1 give large errors when used to predict α (Colwell 1979). Better accuracy is obtained using empirical correlations with fitted parameters – Table 3.2.

Table 3.1. *Theoretical expressions for dispersion density on trays*

Azbel (1963)	$\eta = Fr^{0.5}$
Takahashi *et al.* (1973)	$\eta = 4.1\, Fr^{0.5}$
Kawagoe *et al.* (1976)	$\eta = 1.6\, Fr^{0.33} + 0.22\, Fr$
Kim (1966)	$\eta = (R\, Fr/2)^{0.33}$

$$R = 1 + \frac{Fr}{6} + 0.79\left[\frac{Fr}{2} + \frac{Fr^2}{6} + \frac{Fr^3}{108} + \left(\frac{Fr^2}{4} + \frac{Fr^3}{54}\right)^{0.5}\right]^{0.33} - 0.79\left[\frac{Fr}{2} + \frac{Fr^2}{6} + \frac{Fr^3}{108} - \left(\frac{Fr^2}{4} + \frac{Fr^3}{54}\right)^{0.5}\right]^{0.33}$$

As many of the correlations for α depend on h_{cl}, their detailed comparison is deferred until correlations for h_{cl} have been considered. Whereas most of the correlations in Table 3.2 are based on limited data, those of Colwell (1979) and Stichlmair (1978) are notable exceptions. Colwell took the theoretical expressions of Table 3.1 as a basis and developed his correlation

Table 3.2. *Empirical correlations for dispersion density on sieve and bubble cap trays*

The units are defined in the nomenclature section

Correlation	Conditions
Crozier (1956) $\alpha = \exp[-(0.586\,F_s + 0.45)]$	Bubble cap
Andrew (1969) $\alpha = 1 - (u_s/5.6)^{0.17}$	Sieve, spray regime
Gardner & McLean (1969) $\alpha = 1 - 0.205\,F_s + \dfrac{8.7\,\sigma\rho_L}{\rho_{H2O}} - 0.55$	Sieve, $0.49 < F_s < 1.1$
Kastanek (1970) $\dfrac{(1-\alpha)^3}{\alpha} = 4.45\,u_s$	Bubble cap and sieve
Stichlmair (1978) $\alpha = 1 - (F_s/F_{smax})^{0.28}$	Bubble cap and sieve, $0.03 < \dfrac{F_s}{F_{smax}} < 0.9$
Takahashi *et al.* (1979)	Sieve $d_h < 2$ mm, $u_s < 1.0$ m s^{-1}
$\eta = 8.5\,Fr^{0.5}$	$Fr \leqslant 4.68 \times 10^{-4}\phi^{-0.56}$
$\eta = 1.25\,\phi^{-0.14}\,Fr^{0.25}$	$Fr > 4.68 \times 10^{-4}\phi^{-0.56}$
Colwell (1979) $\eta = 12.6(Fr')^{0.4}\phi^{-0.25}$	Sieve, froth regime
Hofhuis & Zuiderweg (1979)	Sieve
$\eta = 3.4\,Fr^{0.75}$	$Fr < 1.0$ all systems
$\eta = 3.4\,Fr^{0.3}$	$1.0 < Fr < 15.3$ air–water
	$1.0 < Fr < 4.0$ hydrocarbons
$\eta = 420\,Fr\,\rho_G/\rho_L$	$Fr > 15.3$ air–water
	$Fr > 4.0$ hydrocarbons

$F_{smax} = 2.5(\phi^2\sigma(\rho_L - \rho_G)g)^{0.25}$
$Fr' = Fr\,\rho_G/(\rho_L - \rho_G)$

using carefully evaluated data from a wide range of systems operating in the froth regime. Stichlmair used the fluidisation model of spray as the basis for his correlation although he found that it also held in the froth regime. Colwell's correlation is recommended for predicting α in the froth regime and that of Stichlmair for spray.

3.4 Prediction of clear liquid height

3.4.1 *Francis's equation for flow over the exit weir*

Consider a rectangular cross flow tray operating in the froth regime – Fig. 3.2. Bernoulli's equation between sections 1 and 2 for froth flowing at a depth y below the free surface gives

$$\frac{u_{f2}^2 - u_{f1}^2}{2g} + \frac{P_2 - P_1}{\rho_F g} = 0 \tag{3.11}$$

where turbulent flow, an absence of friction and a uniform froth density have been assumed. If the approach velocity is small, $u_{f1} \ll u_{f2}$. The static pressure in the jet as it flows over the weir is equal to the surrounding free surface pressure P_s and also $P_1 = P_s + y\rho_F g$. Substituting in eqn. (3.11) gives the familiar equation

$$u_{f2} = (2gy)^{0.5} \tag{3.12}$$

The volumetric flow of froth over the weir is obtained by integrating the flow through an element of area $W\,dy$ to give

$$Q_F = \int_0^{h_{ow}} C_d W\,dy(2gy)^{0.5} = \frac{2}{3} C_d W (2g)^{0.5} (h_{ow})^{1.5} \tag{3.13}$$

where C_d is introduced to account for friction.

But $Q_L = (1 - \varepsilon_w) Q_F$, where ε_w is the gas holdup in the froth flowing over the weir so that finally we obtain Francis's equation (Francis 1883):

$$h_{ow} = \frac{1.04}{C_d^{0.67} g^{0.33}} \left(\frac{Q_L}{(1 - \varepsilon_w) W} \right)^{0.67} \tag{3.14}$$

Fig. 3.2. Flow of froth over the exit weir.

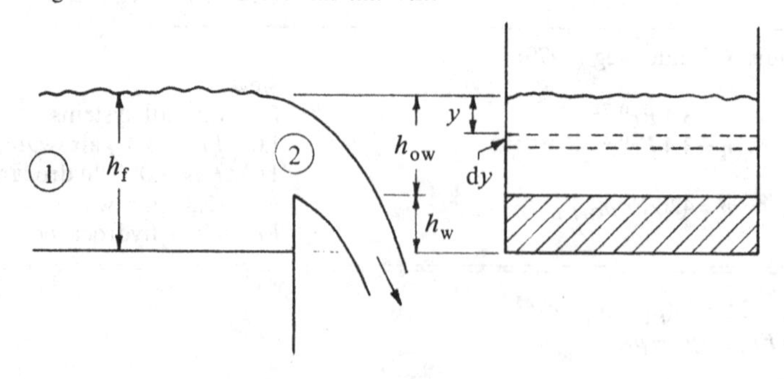

Eqn. (3.14) has been verified for the emulsion regime by Hofhuis & Zuiderweg (1979) who took $C_d = 1.0$ and also by Lockett & Gharani (1979) using $C_d = 0.6$ and a measured value of $\varepsilon_w = 0.5$. Some authors (Unno & Inoue 1976, Darton 1979, Hofhuis 1980) have elaborated on the derivation of eqn. (3.14) taking into account the vertical froth density profile flowing over the weir.

3.4.2 Clear liquid height on sieve trays

3.4.2.1 *Correlations based on Francis's equation*

An estimate of the clear liquid height on the tray is simply obtained from eqn. (3.14) as

$$h_{cl} = (1-\varepsilon)h_f = (1-\varepsilon)(h_w + h_{ow}) \tag{3.15}$$

$$h_{cl} = (1-\varepsilon)\left[h_w + \frac{0.49}{C_d^{0.67}}\left(\frac{Q_L}{(1-\varepsilon_w)W}\right)^{0.67}\right] \tag{3.16}$$

One difficulty in using eqn. (3.16) is that ε_w is unknown. From the liquid holdup profiles of Fig. 2.2, we expect ε_w to be larger than the gas holdup averaged over the total dispersion height ε. So that, although it is convenient to put $\varepsilon_w = \varepsilon$ in eqn. (3.16), this leads to an underestimation of h_{cl}. This approach was adopted by Colwell (1979), who proposed

$$h_{cl} = (1-\varepsilon)\left[h_w + \frac{0.49\,k}{C_d^{0.67}}\left(\frac{Q_L}{(1-\varepsilon)W}\right)^{0.67}\right] \tag{3.17}$$

Colwell verified eqn. (3.17) for the froth regime ($u_s < 2.0\ \mathrm{m\ s^{-1}}$), with $k = 1.49$, using data from a wide variety of systems. The weir coefficient C_d was taken from studies on single-phase flow as follows:

$$C_d = 0.61 + 0.08\frac{h_{ow}}{h_w} \quad \text{when } \frac{h_{ow}}{h_w} \leqslant 8.14 \tag{3.18}$$

and

$$C_d = 1.06\left(1 + \frac{h_w}{h_{ow}}\right)^{1.5} \quad \text{when } \frac{h_{ow}}{h_w} > 8.14 \tag{3.19}$$

where

$$h_{ow} = \frac{h_{cl}}{1-\varepsilon} - h_w$$

As Colwell's correlation for ε, Table 3.2, involves h_{cl}, an iterative calculation is required to determine h_{cl} and ε. A further difficulty is that in the mixed-froth and spray regimes some of the liquid is projected over the weir as drops rather than flowing over as a liquid phase continuous dispersion. This also tends to increase h_{cl} above the value given by eqn. (3.16) and the correction factor k in eqn. (3.17) also includes this effect. However, as Colwell restricted his correlation to the froth regime, the contribution of projected drops was probably small. Stichlmair (1978) adopted a different approach to account for projected drops by adding an additional term to eqn. (3.16). His equation can be written as

$$h_{cl}=(1-\varepsilon)\left[h_w+\frac{0.49}{C_d^{0.67}}\left(\frac{Q_L}{(1-\varepsilon)W}\right)^{0.67}+\frac{125(u_s-u_b)^2\rho_G}{g(\rho_L-\rho_G)\varepsilon^2}\right] \quad (3.20)$$

where

$$u_b=1.55\left(\frac{\sigma(\rho_L-\rho_G)g}{\rho_L^2}\right)^{0.25}\cdot\left(\frac{\rho_G}{\rho_L}\right)^{1/24} \quad (3.21)$$

and $C_d=0.61$. Stichlmair argued that only gas velocities in excess of the single bubble rise velocity u_b are significant in causing drop projection over the weir.

When drop projection is significant, in the spray and mixed-froth regimes, Hofhuis & Zuiderweg (1979) proposed for the clear liquid height flowing over the weir

$$(h_{cl})_{ow}=0.26\,\psi^{-0.37}\left(\frac{Q_L}{W}\right)^{0.67} \quad (3.22)$$

where

$$\psi=\frac{Q_L}{Wu_s}\cdot\left(\frac{\rho_L}{\rho_G}\right)^{0.5} \quad (3.22a)$$

Another recent correlation due to Bennett *et al.* (1983), also based on eqn. (3.16), is

$$h_{cl}=\alpha_e\left[h_w+C\left(\frac{Q_L}{W\alpha_e}\right)^{0.67}\right] \quad (3.23)$$

where

$$\alpha_e=\exp\left[-12.55\left(u_s\left(\frac{\rho_G}{\rho_L-\rho_G}\right)^{0.5}\right)^{0.91}\right] \quad (3.24)$$

and

$$C=0.50+0.438\exp(-137.8\,h_w) \quad (3.25)$$

3.4.2.2 *Other empirical correlations for h_{cl} on sieve and bubble cap trays*

A large number of empirical correlations have been proposed for h_{cl} which are not based on Francis's equation. Hofhuis & Zuiderweg (1979), for example, have proposed for both froth and spray:

$$h_{cl}=0.6(\psi p)^{0.25}h_w^{0.5} \quad 0.007\,\text{m}<d_h<0.01\,\text{m} \quad (3.26)$$

Eqn. (3.22*a*) is used for ψ. The validity of eqn. (3.26) in the spray regime is questionable since the available experimental evidence (Fane & Sawistowski 1969, Pinczewski & Fell 1972, Lockett 1981) suggests that h_{cl} is almost independent of the weir height in spray. Even in the froth regime, it is clearly suspect as either h_w or Q_L approach zero. Many proposed correlations are of the form

$$h_{clD} = \alpha' h_w - \beta F_s + \gamma \frac{Q_L}{W} + \delta \tag{3.27}$$

where the units are defined in the Nomenclature Section. A summary of these correlations is given in Table 3.3. Eqn. (3.7) is used to obtain h_{cl} from h_{clD}.

3.5 Prediction of dispersion height

In view of eqn. (3.1), dispersion height, h_f, can be predicted by combining correlations for h_{cl} and α. Alternatively, numerous correlations have been proposed of the form of eqn. (3.28) and they are also summarised in Table 3.3:

$$h_f = Ah_w + BF_s + C\frac{Q_L}{W} + D \tag{3.28}$$

3.6 Which correlations to use?

Some of the proposed correlations for α, h_{cl} and h_f are compared in Figs. 3.3–3.5. Columns having small diameters, excessively long exit calming zones or baffles over the exit weir tend to have increased clear liquid and dispersion height. Colwell, in particular, excluded such data in developing his correlations. These effects are discussed below. Some of the more recent correlations take the flow regime into account and these are to be preferred for that reason. Also correlations which are based on a wide range of data from many sources are to be preferred over those based on the limited data of a single investigator. On this basis, the correlations of either Stichlmair or Colwell for h_{cl} and α are recommended for the froth regime. Bennett's correlation is also useful for h_{cl}. For the spray regime the situation is more uncertain, although Stichlmair's correlation can perhaps be employed. Note that none of the correlations takes into account bubble or drop stabilisation arising from surface tension gradients as discussed in Section 2.8.2.

3.7 The influence of liquid exit and entry conditions

3.7.1 *Exit calming zones*

An unperforated exit calming zone can be caused by the presence of a tray support beam or by deliberate blanking of part of the tray active area, perhaps because the column is working below normal design loads. An exit calming zone increases both h_f and h_{cl} on the tray.

Table 3.3. *Clear liquid height and dispersion height correlations – parameters in eqns. (3.27) and (3.28)*

Source	α'	β	γ	δ	A	B	C	D	Comments
Gerster *et al.* (1958)	0.29	0.0135	2.45	0.040	0.73	0.067	10.3	0	Bubble caps, air–water
AIChE (1958)					1.89	$0.043\,F_s$	0	−0.041	Bubble caps, organic mixtures
Barker & Self (1962)	0.37	0.012	1.78	0.024	1.06	0.035	4.82	0.035	Sieve, air–water $d_h = 4.8$ mm
Harris & Roper (1962)	0.58	$0.23\,h_w$	3.68	0.006	0.75	0.021	12.3	0.038	Sieve, air–water, $d_h = 4.8$ mm[a]
Gerster, quoted in Smith (1963)	0.725	$0.238\,h_w$	1.23	0.006					Sieve, air–water
Finch & Van Winkle (1964)	0.22	0.012	4.54	0.025	0.74	$(0.039\,F_s - 0.095)$	11.4	0.12	Sieve, air–water–methanol, $d_h = 1.6$–8 mm[b]
Thomas & Campbell (1967)	0.19	0.0083	1.59	0.040	1.24	0.051	5.42	0	Sieve, air-aq. glycerol $d_h = 3.2$ mm, $Z_e = 152$ mm
Brambilla *et al.* (1969)	0.79	0.0123	2.05	0.013					Bubble caps, air–water
Brambilla *et al.* (1969)	0.72	0.025	2.16	0.003					Sieve, air–water $d_h = 5$ mm, $0.4 < F_s < 1.3$
Kastanek (1970)					2.95	$\frac{0.226}{\rho_G^{0.5}}$	0	0.024	Bubble cap, APV West, sieve, total reflux, $h_f^{0.5} = Ah_w + BF_s + D$

Thomas & Haq (1976)	0.42	0.0096	3.93	0.031	0.205	0.067	9.81	0	Sieve, air/CO_2–aq. glycerol, $d_h = 9.6$ mm, $Z_e = 150–190$ mm
Thomas & Haq (1976)					0.43	0.069	11.0	0	Sieve, air/CO_2–aq. Na_2CO_3, $d_h = 9.6$ mm, $Z_e = 150–190$ mm

[a] Column diameter = 0.3 m
[b] Clear liquid height from $h_{WT} - h_{DT}$
Z_e = length of exit calming zone

Fig. 3.3. Comparison of correlations for dispersion density on sieve trays. $Q_L/W = 0.01\ m^3\ m^{-1}\ s^{-1}$, air–water, $h_w = 0.05\ m$, $\phi = 0.1$, $d_h = 0.0127\ m$.

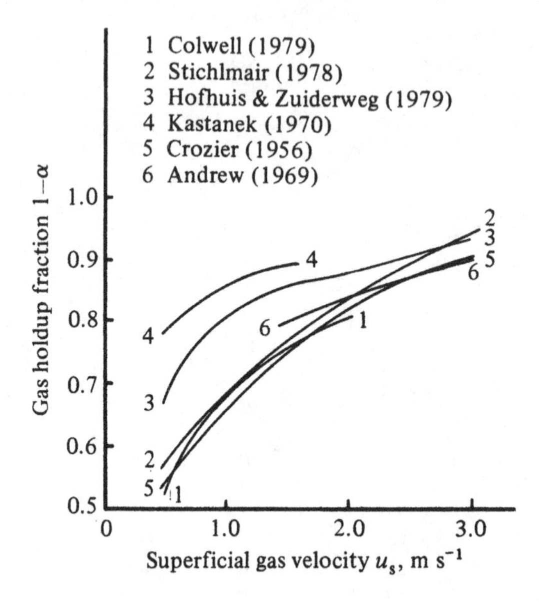

Fig. 3.4. Comparison of correlations for clear liquid height on sieve trays – conditions as for Fig. 3.3.

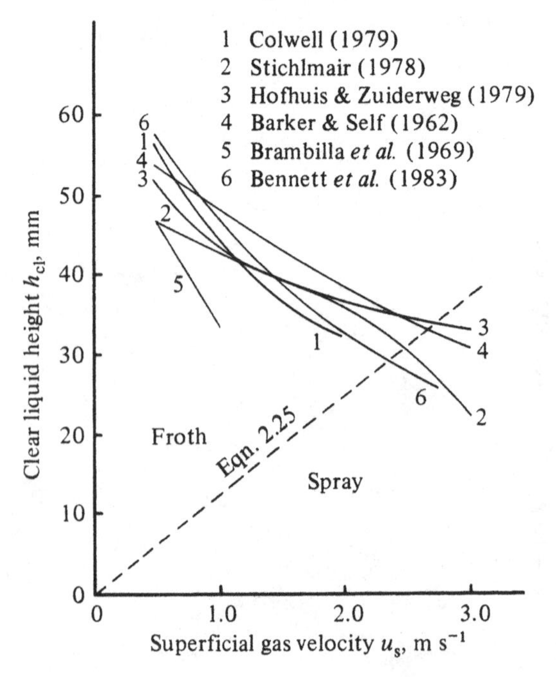

The situation can be modelled by assuming that the liquid flowing over the exit weir is completely degassed ($\alpha = 1$) – Fig. 3.6. The variation of dispersion density with height above the tray is important and as an approximation can be taken as linear in the froth regime. Use of the horizontal momentum equation between sections 1 and 2 is shown in Appendix A to give (Colwell 1979, Lockett & Uddin 1978)

Fig. 3.5. Comparison of correlations for dispersion height on sieve trays – conditions as for Fig. 3.3.

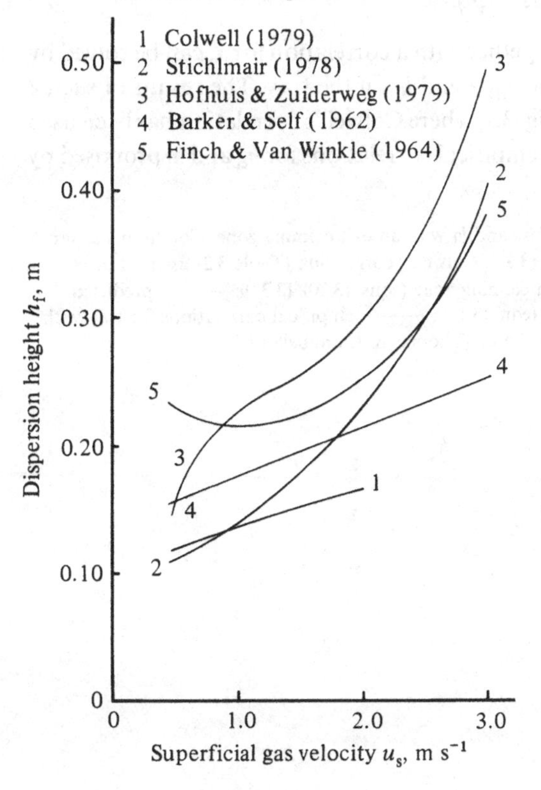

Fig. 3.6. Exit calming zone.

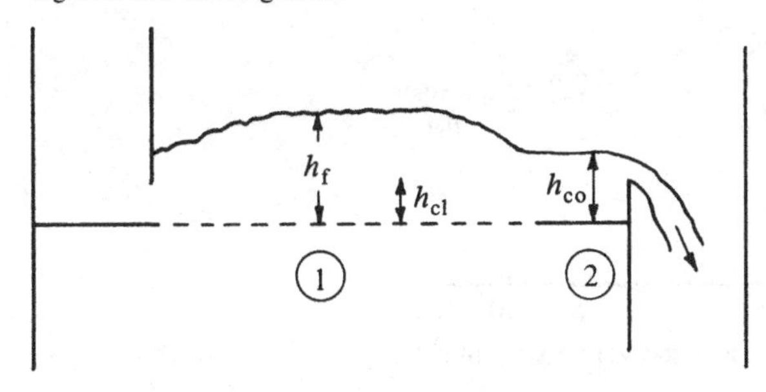

$$\frac{Q_L^2 \rho_L}{h_{cl} W} + \frac{1}{3} h_f^2 W g \alpha \rho_L = \frac{Q_L^2 \rho_L}{h_{co} W} + \frac{1}{2} h_{co}^2 W \rho_L g \tag{3.29}$$

Using eqns. (3.1) and (3.29) we get

$$h_{cl} = h_{co} \left[1.5\alpha - \frac{3Q_L^2 \alpha}{h_{co}^2 W^2 g} \left(\frac{1}{h_{cl}} - \frac{1}{h_{co}} \right) \right]^{0.5} \tag{3.30}$$

where, from eqn. (3.14),

$$h_{co} = h_w + \frac{1.04}{C_d^{0.67} g^{0.33}} \left(\frac{Q_L}{W} \right)^{0.67} \tag{3.31}$$

Eqns. (3.30) and (3.31), together with a correlation for α, can be solved by trial and error to determine h_{cl}, α and h_f on the tray. The results of such a calculation are shown in Fig. 3.7, where Colwell's correlation has been used for α. Also included are the empirical correlations for h_{cl} and h_f proposed by

Fig. 3.7. Prediction of h_{cl} and h_f with an exit calming zone. Conditions as for Fig. 3.3; C_d from eqn. (3.18). Colwell's correlation (Table 3.2) used for α. ——— Predicted with calming zone (eqns. (3.30), (3.31)); — — — predicted without calming zone (eqn. (3.17)); - - - - - empirical correlations for tray with exit calming zone of 152 mm. (Thomas & Campbell 1967.)

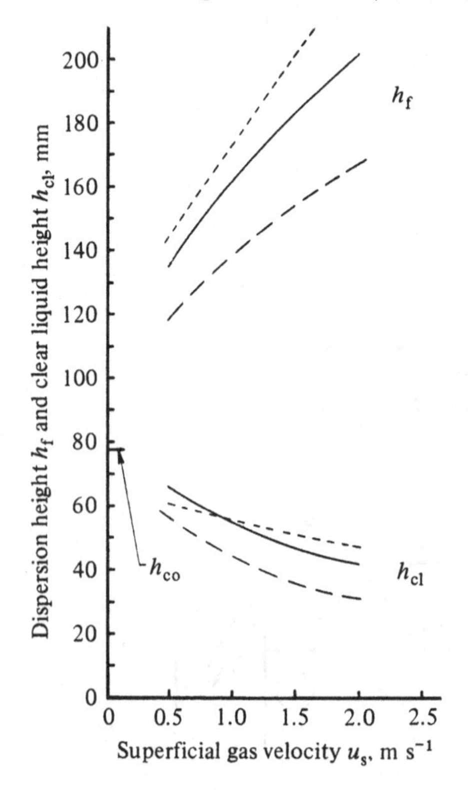

Thomas & Campbell (1967) (Table 3.3) for a sieve tray having an exit calming zone of 152 mm. We can conclude that the fundamental approach represented by eqns. (3.30) and (3.31) agrees very well with Thomas & Campbell's empirical correlations. Also shown on Fig. 3.7 are Colwell's predictions (eqn. (3.17)) in the absence of a calming zone. Note that an exit calming zone increases both h_{cl} and h_f on the tray. Stichlmair (1978) also found experimentally that the froth height on the tray increased as the length of the exit calming zone increased. However, only Colwell appears to have specifically taken the calming zone length into account in developing his h_{cl} correlation. Obviously for short calming zones and high liquid loads, the liquid flowing over the weir is only partly degassed and h_{cl} will then lie between the normal value and that predicted from eqn. (3.30). Note also that the analysis holds only for the froth or emulsion flow regimes. In the spray regime the length of the exit calming zone appears to be unimportant (Lockett 1981).

3.7.2 *Splash baffles*

Since liquid weir load increases with increasing column diameter, the weir loads typically encountered in small-diameter laboratory columns are abnormally low. Consequently, some workers have used a splash baffle over the exit weir (Fig. 3.8) to increase the dispersion height to the level obtained on larger industrial trays (Sargent *et al.* 1964, Haselden & Thorogood 1964, Rush & Stirba 1957). In the presence of a splash baffle, h_{cl} on the tray can be obtained from a pressure balance under the baffle using eqn. (3.14) (Colwell 1979):

$$h_{cl} = h_w + \frac{1.04}{C_d^{0.67} g^{0.33}} \left(\frac{Q_L}{W}\right)^{0.67} + k'k'_{da} \qquad (3.32)$$

Fig. 3.8. Splash baffle.

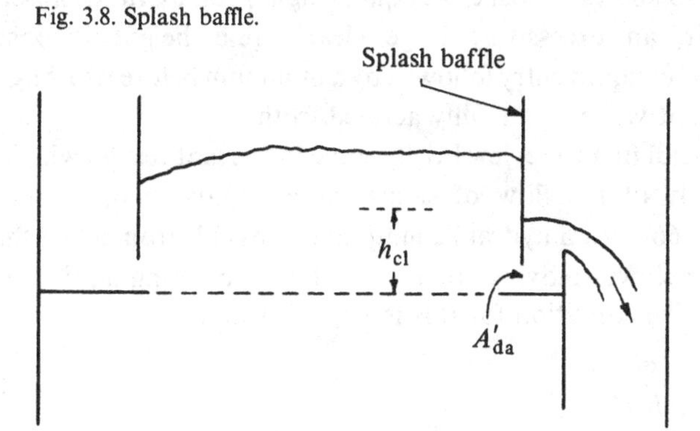

where

$$h'_{da} = 0.165\left(\frac{Q_L}{A'_{da}}\right)^2 \tag{3.33}$$

Eqn. (3.33) gives the head loss for flow under the baffle where A'_{da} is the vertical area available for flow. If the baffle is set too close to the exit weir, the horizontal area between the baffle and the weir also becomes significant. Colwell recommended a value of $k' = 2$ as an average value to allow for this.

3.7.3 *Converging flow over the exit weir*

In small round columns, the liquid converges as it enters the downcomer, so the effective weir length becomes progressively shorter as the liquid flows over the weir. Bolles (1946) published a correction to Francis's equation to allow for the effect. However, his derivation was based on an assumed fixed jet trajectory and considered only a single-phase fluid so it can only be considered approximate. Another problem is that the downcomer width, between the weir and the wall, falls to zero at the ends of the weir. Both these effects increase clear liquid and froth heights in small round columns above those on equivalent rectangular columns at the same loads. Colwell (1979) suggested that these effects need only be considered for column diameters less than 1.2 m, but at present there is no established way of accounting for them.

3.7.4 *Liquid entry effects*

Whereas the froth height on the bulk of the tray is set by conditions downstream, for example by the exit weir and the crest over it, the liquid height immediately on entering the tray from the downcomer is also influenced by an upstream condition – the clearance under the downcomer. There is also a transition from essentially single-phase to two-phase flow and both these factors make analysis in the region of the liquid entry quite difficult. Fig. 3.9 shows typical clear liquid height profiles. In the absence of an inlet weir, an excessively large clear liquid height is observed immediately after liquid entry followed by a minimum before rising again to a fairly constant value in the fully aerated froth.

It will be useful first to review briefly some important results which arise from the study of the flow of single-phase liquids in open channels (Henderson 1966). For a hydraulic jump to be possible from a fast shallow jet to a deeper slower-moving stream, the flow of entering liquid must be supercritical. The condition for this is $Fr_1 > 1$, where

$$Fr_1 = \frac{u_{L1}}{(gh_1)^{0.5}} \tag{3.34}$$

If we neglect jet contraction and assume the liquid is unaerated as it leaves the downcomer of clearance h_1, this implies that for a hydraulic jump to be possible

$$h_1 < \left(\frac{Q_L}{Wg^{0.5}}\right)^{0.67} \tag{3.35}$$

If a hydraulic jump occurs, the liquid increases from a depth h_1 to its conjugate depth h_2, where the two are related by the momentum equation

$$\frac{Q_L^2 \rho_L}{h_1 W} + \frac{1}{2} h_1^2 W \rho_L g = \frac{Q_L^2 \rho_L}{h_2 W} + \frac{1}{2} h_2^2 W \rho_L g \tag{3.36}$$

One way to get around the problem of dealing with both single-phase and two-phase fluids in combination is to realise that eqn. (3.29) relates a two-phase mixture of height h_f and liquid holdup fraction α to a depth h_{co} of clear liquid through the momentum equation. It follows that when dealing with the momentum equation, as we are here, we can replace the two-phase mixture by a single-phase liquid of depth h_{co} calculated by trial and error from eqn. (3.37):

$$h_{co} = h_{cl}\left[1.5\alpha - 3\left(\frac{Q_L}{W}\right)^2 \frac{\alpha}{g} \frac{1}{h_{co}^2}\left(\frac{1}{h_{cl}} - \frac{1}{h_{co}}\right)\right]^{-0.5} \tag{3.37}$$

Fig. 3.10 shows four alternatives which can exist at liquid entry. Fig. 3.10*a* shows a drowned subcritical entry. Here $Fr_1 < 1$ as the liquid issues from the downcomer, so the flow is subcritical. In addition, the initial depth h_1 is less than h_{co} corresponding to liquid downstream. This causes the entering

Fig. 3.9. Clear liquid height profiles on a sieve tray (Dhulesia 1980). $u_s = 1.9\ \mathrm{m\,s^{-1}}$, $Q_L/W = 0.017\ \mathrm{m^3\,m^{-1}\,s^{-1}}$, $h_w = 0.05$ m, $d_h = 12.7$ mm, $\phi = 0.1$, downcomer clearance 0.05 m, air–water.

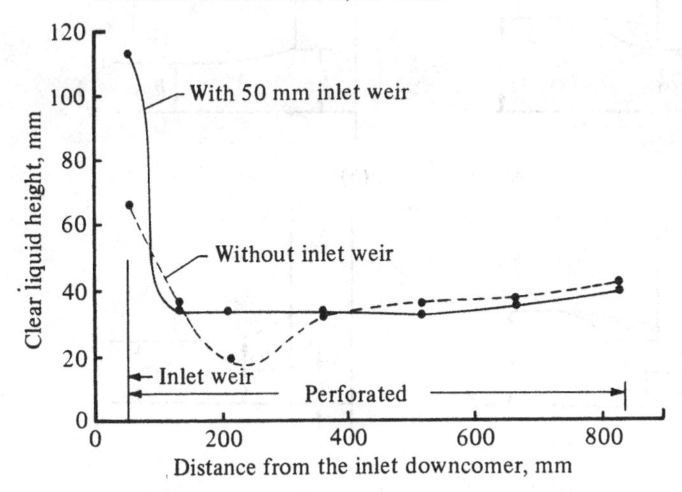

liquid stream to be drowned by a pool of liquid (or in reality a two-phase mixture), which is relatively stagnant compared with the underflowing liquid stream. Fig. 3.10*b* shows a subcritical liquid entry which is not drowned. Here the depth of the entering liquid is greater than h_{co}. The downcomer is assumed to remain sealed by froth. Fig. 3.10*c* shows a drowned supercritical entry. Although the entering liquid is supercritical and the possibility of a hydraulic jump exists, the conjugate depth h_2 is less than h_{co} so the entering jet becomes drowned. Fig. 3.10*d* shows a case of supercritical entry with a hydraulic jump. As the entering jet slows down by friction and aeration, a point is reached where its conjugate depth becomes equal to h_{co} and a hydraulic jump occurs.

It normally turns out that the clearance under the downcomer is set such that the entering jet is drowned. Fig. 3.11 shows how a drowned entry can be approximately analysed. It is assumed that just after entry both the jet and the stagnant pool of liquid overlaying it are unaerated. A momentum balance between the inlet and the fully aerated froth then gives

$$\frac{Q_L^2 \rho_L}{h_1 W} + \frac{1}{2} h_i^2 W \rho_L g = \frac{Q_L^2 \rho_L}{h_{cl} W} + \frac{\alpha \rho_L g W h_f^2}{3} \tag{3.38}$$

Fig. 3.10. Alternative liquid entry conditions. (*a*) Subcritical drowned entry: $Fr_1 < 1$, $h_{co} > h_1$. (*b*) Subcritical non-drowned entry: $Fr_1 < 1$, $h_1 > h_{co}$. (*c*) Supercritical drowned entry: $Fr_1 > 1$, $h_{co} > h_2$. (*d*) Supercritical hydraulic jump entry: $Fr_1 > 1$, $h_2 > h_{co}$.

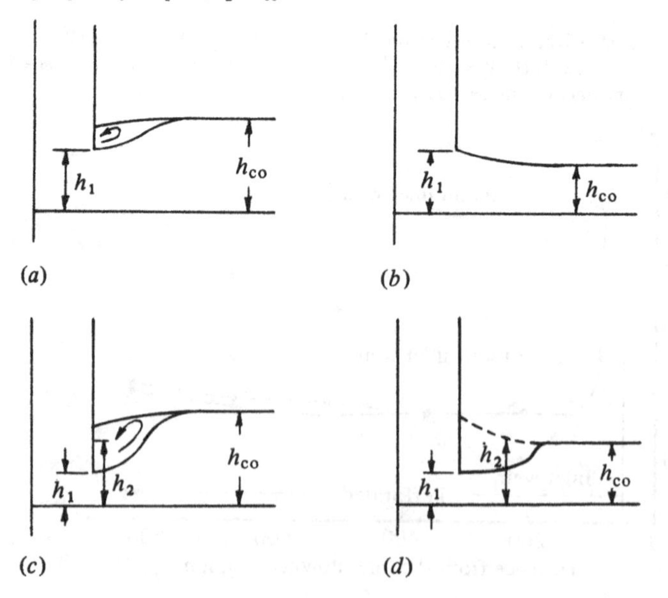

from which it follows that:

$$h_i = \left[\frac{2}{g}\left(\frac{Q_L}{W}\right)^2\left(\frac{1}{h_{cl}} - \frac{1}{h_1}\right) + \frac{2\alpha h_f^2}{3}\right]^{0.5} \tag{3.39}$$

The value of h_i is important for calculating the backup of froth in the downcomer (Section 5.3.1).

For the situation shown in Fig. 3.9 in the absence of an inlet weir, $h_{cl} =$ 0.035 m and $\alpha = 0.17$ using Stichlmair's equation in Table 3.2. Since $Fr_1 =$ 0.49 and $h_{co} = 0.075$ m from eqn. (3.37), it follows that since the downcomer clearance h_1 is 0.05 m this is a case of a drowned subcritical entry. Using eqn. (3.39), $h_i = 0.073$ m, which is in reasonable agreement with the measured value shown in Fig. 3.9.

The theory given above has not yet been tested against experiment to any extent. It can be refined by taking into account the contraction of the jet flowing under the downcomer, and Khamdi *et al.* (1963) have suggested a contraction coefficient of 0.62 for this. Furthermore, the theory gives no information about the location or magnitude of the observed minimum in the clear liquid height just downstream of liquid entry. The theory does appear to be an improvement over previous analyses. For example, the analysis of Khamdi *et al.* (1963) considers only single-phase flow. Stichlmair (1978) suggested that a hydraulic jump could be avoided if $\rho_L u_{L1}^2 < u_h^2 \rho_G$. Both approaches are obviously oversimplified.

Hydraulic jumps can be reduced and smoother entry conditions obtained by the use of an inlet weir as shown in Fig. 3.9. A recessed receiving pan under the downcomer can serve the same purpose. However, receiving pans are expensive and unsuitable for dirty services. At high liquid flow rates an inlet weir can cause liquid to be projected over it to land some distance downstream where the impact momentum can cause weeping. A small inlet weir or interrupter bar is often used on valve trays to discourage the entering liquid jet from hitting the first row of valves where it can cause excessive weeping.

Fig. 3.11. Drowned liquid entry.

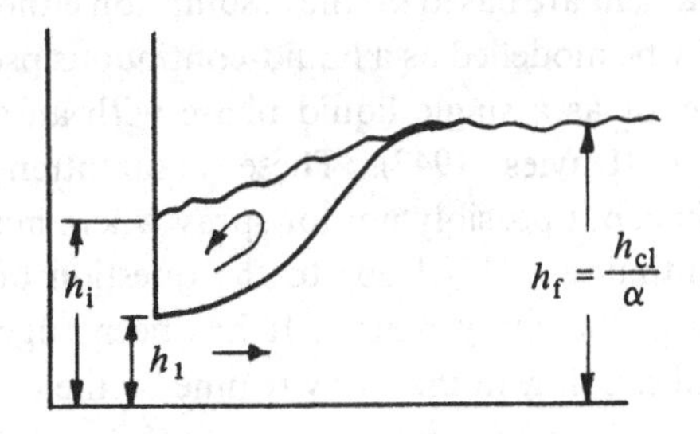

The clear liquid height in the liquid entry region tends to be higher than on the rest of the tray. This results in a reduced vapour flow through the first few rows of holes since the wet tray pressure drop is constant over all parts of the tray. The result is an effective loss of bubbling area and a tendency to weep at liquid entry, which both reduce tray efficiency. Two designs of 'bubble promoter' have been proposed (Smith & Delnicki 1975, Winter & Uitti 1976). These involve raising the perforated tray deck by a few millimetres in the liquid entry region. By so doing, preferential vapour flow is encouraged which helps minimise the adverse effects discussed above.

3.8 Hydraulic gradient

As liquid flows across the tray, it experiences a resistance from the tray hardware and from the vapour passing through it. The hydraulic gradient is the fall in clear liquid height per unit tray length arising from these resistances. Hydraulic gradient is considerable on bubble cap trays and calculation methods are well established (Davies 1947, Bolles 1956). It is less important on sieve trays and valve trays although data is lacking in the literature for the latter. Of the sieve tray studies (Mersmann 1963, Sterbacek 1967, Hughmark & O'Connell 1957, Fair 1963), Fair's correlation can conveniently be represented as

$$\text{hydraulic gradient } \frac{\Delta h_{cl}}{Z} = \frac{f u_f^2}{g R_h} \tag{3.40}$$

where friction factor

$$f = 7 \times 10^4 \, h_w Re_f^{-1.06} \tag{3.41}$$

and

$$Re_f = \frac{R_h u_f \rho_L}{\mu_L}, \quad u_f = \frac{Q_L}{W' h_{cl}}, \quad R_h = \frac{W' h_f}{W' + 2h_f} \tag{3.42}$$

W' = average flow-path width for liquid flow $= (W+D)/2$ for single-pass trays. The significance of the weir height in eqn. (3.41) may be that as h_w increases, the dispersion height and vapour residence time also increase and increasing the latter increases the resistance to liquid flow. The available correlations for hydraulic gradient are based on the assumption either that the two-phase dispersion may be modelled as a liquid-continuous pseudo-fluid as in eqn. (3.40) above, or as a single liquid phase with an added correction for vapour flow (Davies 1947). These assumptions are reasonable for the froth regime, but possibly not for spray where much of the dispersion is vapour-continuous. This leads to the question of how liquid flows across the tray in the spray regime. It has been suggested (Porter *et al.* 1977) that liquid can flow in the spray regime by the random movement of the liquid drops if there is a high drop concentration at the tray inlet which falls as the drops cross the tray. However, the measured rates of liquid mixing in the spray regime (Section 9.7) are insufficient to transport more than a very low liquid flow rate by this mechanism. Consequently, liquid flow in the spray regime is believed to consist of two parts. The bulk of the liquid flows as a liquid-continuous phase near the tray floor and above this the remainder of the liquid flows by random movement of liquid drops with a net flow of drops towards the exit weir (Manickampillai & Sawistowski 1981).

Based on the above picture of liquid flow in the spray regime, which is not radically different from that in the froth regime, we might expect a single correlation, eqn. (3.40), to be adequate for predicting hydraulic gradients in both regimes and this proves to be the case.

The hydraulic gradient increases the clear liquid and dispersion heights progressively upstream from the exit weir. These effects are largest at high liquid loads and for long liquid flow paths. Hydraulic gradient is usually only significant on sieve trays having a low pressure drop and long liquid flow path, such as those used in vacuum distillation. Then it can cause vapour maldistribution or excessive weeping near the liquid inlet. Where such a possibility is recognised as the design stage, the usual practice is to increase the number of passes to reduce the liquid flow path length or to increase the dry tray pressure drop so as to reduce weeping and vapour maldistribution. In fact, because of entry and exit effects for small to medium flow path lengths, the hydraulic gradient can turn out to be negative as shown in Fig. 3.9.

3.9 Clear liquid height on valve trays

It is more difficult to measure h_{cl} on valve than on sieve trays. The use of γ-ray absorption is inaccurate near the tray floor because of the

Fig. 3.12. Showing why a manometer overestimates h_{cl} on a valve tray.

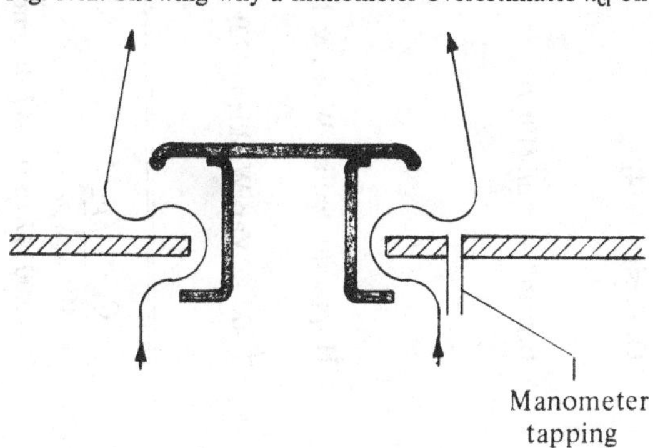

Table 3.4. *Clear liquid height and froth height correlations for valve trays*

Author	Correlation	Comments
Dhulesia (1984)	$h_{cl}=0.42\,\psi^{0.33}h_w^{0.67}$	Glitsch V1 valves, air–water $0.025<h_w<0.075$ m $0.03<\psi<0.3$ m
Brambilla *et al.* (1969)	$h_{clD}=0.74\,h_w-0.0145\,F_s+1.66\frac{Q_L}{W}+X$	Air–water $0.03<h_w<0.07$ m $0.4<F_s<1.3$ $1.4\times10^{-3}<\frac{Q_L}{W}<7.5\times10^{-3}$
Hoppe quoted by Weiss & Langer (1979)	$h_{cl}=0.982\,\phi^{-0.25}u_h^{-0.365}h_w^{0.575}\left(\frac{Q_L}{W}\right)^{0.235}$	
Todd & Van Winkle (1972)	$h_f=0.076+\frac{32.6\,F_s^2}{(\rho_L-\rho_G)}+0.82\,h_{clD}$	For distillation systems $D=0.46$ m

$\psi=\frac{Q_L}{W}\cdot\frac{1}{u_s}\left(\frac{\rho_L}{\rho_G}\right)^{0.5}$

$X=0.00147$ to 0.0047 depending on valve type

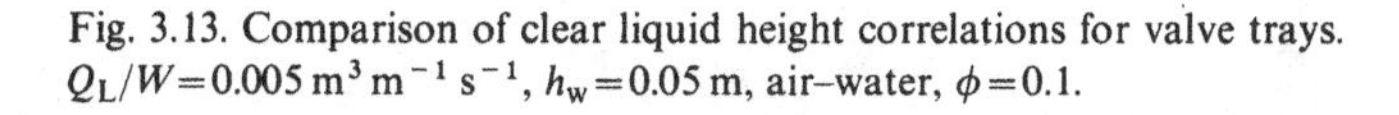
Fig. 3.13. Comparison of clear liquid height correlations for valve trays. $Q_L/W = 0.005\ m^3\ m^{-1}\ s^{-1}$, $h_w = 0.05$ m, air–water, $\phi = 0.1$.

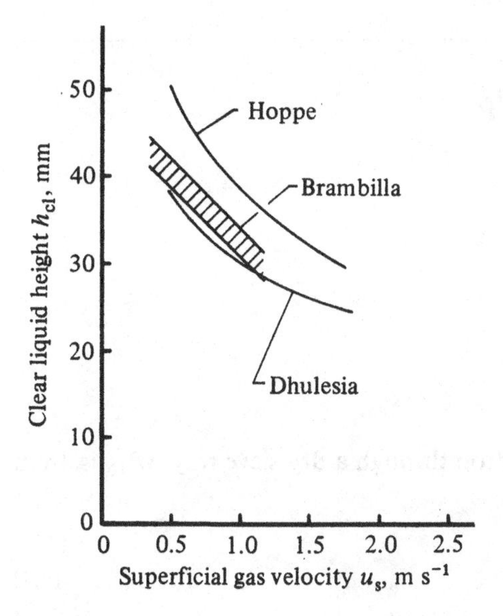

presence of the valves. Also the manometric technique, described in Section 3.2, usually overestimates h_{cl} because for most valve shapes the vapour is deflected by the valve slightly towards the tray floor; see Fig. 3.12. A manometer set flush with the tray measures not only the liquid head, but also the impact pressure of the vapour. Dhulesia (1984) used a baffle around the manometer tapping in an effort to overcome this problem.

Some proposed correlations for h_{cl} on valve trays are summarised in Table 3.4 and compared in Fig. 3.13. Other proposed correlations (Piqueur & Verhoeye 1976, Todd & Van Winkle 1972, Weiss & Langer 1979) contain obvious errors. The correlations of Dhulesia and Brambilla are in reasonable agreement and are recommended. Also included in Table 3.4 is a correlation for h_f on valve trays proposed by Todd & Van Winkle.

There is a definite need for a new correlation based on a realistic model for liquid flow, similar to eqn. (3.16).

4

Pressure drop

4.1 Basic equations

The vapour pressure drop through a dry sieve tray ΔP_{13}^{d} is, from Fig. 4.1*a*,

$$\Delta P_{13}^{d} = \Delta P_{12}^{d} + \Delta P_{23}^{d} \tag{4.1}$$

where $\Delta P_{13}^{d} = P_1^{d} - P_3^{d}$, etc. Uniform pressure is assumed at each horizontal control surface. The pressure recovery ΔP_{23}^{d} is negative. The wet tray pressure drop for the same vapour flow through a tray carrying liquid is, from Fig. 4.1*b*,

$$\Delta P_{13}^{w} = \Delta P_{12}^{w} + \Delta P_{23}^{w} \tag{4.2}$$

The pressure drop through the tray deck is altered by the presence of liquid and

$$\Delta P_{12}^{w} = \Delta P_{12}^{d} + \Delta P_{R1} \tag{4.3}$$

ΔP_{R1} arises because of: surface tension forces; the presence of liquid may affect the way the vapour flows into the holes so altering the discharge

Fig. 4.1. Control surfaces for pressure drop denoted by ----. (*a*) Without liquid. (*b*) With liquid.

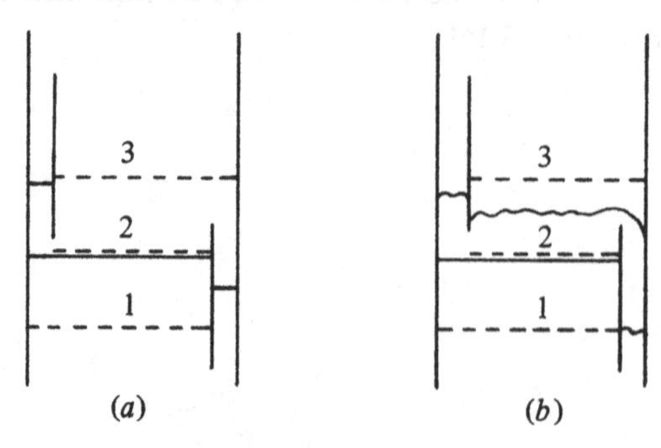

coefficient; some of the holes may be partially blocked by liquid; and variations in the local liquid head may cause local fluctuations in the vapour flow, and a fluctuating vapour flow has a larger pressure drop than an equal steady flow (Sargent *et al.* 1964, Davies & Porter 1965).

The dynamic and static pressure drops over the liquid are related by eqn. (3.7), so that

$$\Delta P^{\mathrm{w}}_{23} = h_{\mathrm{cl}}\rho_{\mathrm{L}}g - u_{\mathrm{s}}\rho_{\mathrm{G}}(u_{\mathrm{h}} - u_{\mathrm{s}}) + \Delta P_{\mathrm{R2}} \tag{4.4}$$

ΔP_{R2} is included because eqn. (3.7) is approximate and neglects both pressure variations along the tray floor and also the vena-contracta. The latter can cause the vapour velocity leaving the holes to be larger than u_{h} since u_{h} is based on the cross-sectional area of the holes.

Davies & Porter (1965) found complete pressure recovery above a dry tray having $t/d_{\mathrm{h}} = 1$, i.e. from a momentum balance between 2 and 3 on Fig. 4.1*a*,

$$-\Delta P^{\mathrm{d}}_{23} = u_{\mathrm{s}}\rho_{\mathrm{G}}(u_{\mathrm{h}} - u_{\mathrm{s}}) + \Delta P_{\mathrm{R3}} \tag{4.5}$$

ΔP_{R3} is included because eqn. (4.5) has only been verified for one particular tray geometry and it is doubtful whether it is generally valid. From eqns. (4.1)–(4.5)

$$\Delta P^{\mathrm{w}}_{13} = \Delta P^{\mathrm{d}}_{13} + h_{\mathrm{cl}}\rho_{\mathrm{L}}g + \Delta P_{\mathrm{R}} \tag{4.6}$$

where

$$\Delta P_{\mathrm{R}} = \Delta P_{\mathrm{R1}} + \Delta P_{\mathrm{R2}} + \Delta P_{\mathrm{R3}}$$

In terms of liquid head, eqn. (4.6) is

$$h_{\mathrm{WT}} = h_{\mathrm{DT}} + h_{\mathrm{cl}} + h_{\mathrm{R}} \tag{4.7}$$

Correlations for predicting h_{cl} have been given in Chapter 3 and we now consider the other terms in eqn. (4.7).

4.2 Dry tray pressure drop – sieve trays

4.2.1 *Orifice coefficients*

The usual practice is to represent h_{DT} by an orifice-type equation:

$$h_{\mathrm{DT}} = \frac{\xi\rho_{\mathrm{G}}u_{\mathrm{h}}^2}{2g\rho_{\mathrm{L}}} \tag{4.8}$$

(Note that the discharge coefficient $C_{\mathrm{d}} = \xi^{-0.5}$.)

There are over 20 correlations available in the literature for the orifice coefficient (Arnold *et al.* 1952, Mayfield *et al.* 1952, Kamei 1954, Hunt *et al.* 1955, Jones & Pyle 1955, Leibson *et al.* 1956, Hughmark & O'Connell 1957, Kolodzie & Van Winkle 1957, Eduljee 1958*a*, Huang & Hodson 1958, McAllister *et al.* 1958, Mukhlenov & Tarat 1958, Smith & Van Winkle 1958, Prince 1960, Molokanov 1963, Bene 1964, Sterbacek 1964, Zelfel

1965, 1967, Kneule & Zelfel 1966, Muhle 1972, Cervenka and Kolar 1973*b*, Stichlmair & Mersmann 1978).

Most are based on a rather limited amount of data. As for single orifices, ξ is a function of the hole Reynolds number. It is also influenced by the interaction of the vapour jets as they issue from the holes and so depends on ϕ or equivalently on p/d_h. The biggest influence on ξ, however, is provided by the ratio of tray thickness to hole diameter. Huesmann (1966) argued that for thin trays, $t/d_h \rightarrow 0$, the vena-contracta of the vapour jet lies outside and above the tray, and there is a pressure recovery as the vapour slows down from its maximum to its superficial velocity. For thick trays, $t/d_h \gg 0$, the vena-contracta lies within the hole itself. The jet partly expands to fill the complete cross-section of the hole before emerging, after which it slows down to the superficial velocity. The pressure recovery associated with a two-stage expansion is greater than for a single-stage expansion so that thick trays generally have a lower pressure drop than thin trays.

Alternatively, for a fixed tray thickness, the dry tray pressure drop decreases as the hole diameter is decreased while keeping ϕ constant. This is, of course, contrary to what one might expect from simple considerations of pressure drop arising from frictional resistance of the vapour with the sides of the hole. It also does not hold for very small holes or very thick trays, where $t/d_h > 2$, as then a significant pressure drop occurs as the vapour flows along the 'pipe' formed by the hole.

Fig. 4.2 shows a correlation for ξ_0 proposed by Stichlmair & Mersmann (1978). To correct for fractional perforated area they proposed that

$$\xi = \xi_0 + \phi^2 - 2\phi\xi_0^{0.5} \quad \text{for } t/d_h < 2$$

and

$$\xi = \xi_0 + \phi^2 - 2\phi \qquad \text{for } t/d_h > 2 \tag{4.9}$$

A widely used correlation, which differs only slightly in practice from the one above, is that of Smith and Van Winkle (1958), where K_w is determined from Fig. 4.3, and

$$\xi = \frac{(1-\phi^2)(p/d_h)^{0.2}}{K_w^2} \tag{4.10}$$

Cervenka & Kolar (1973*b*) concluded that the effect of Re_h on ξ was insignificant for thin trays and they proposed

$$\xi = \frac{A(1-\phi^2)}{\phi^{0.2}(t/d_h)^{0.2}} \qquad \begin{array}{l} 0.1 < t/d_h < 0.8 \\ 0.015 < \phi < 0.2 \end{array} \tag{4.11}$$

($A = 0.94$ or 1.0 for holes on triangular and square pitch, respectively.) Eqn. (4.11) is convenient because of its analytical form, and it was reported to give an average deviation of 5.4% between experimental and calculated values of ξ.

The correlations above are for drilled holes. Punched holes give a slightly lower pressure drop, particularly if the punch entry side of the tray faces down (upstream).

Fig. 4.2. Orifice coefficient correlation due to Stichlmair & Mersmann (1978). Parameter $Re_h = u_h \rho_G d_h / \mu_G$.

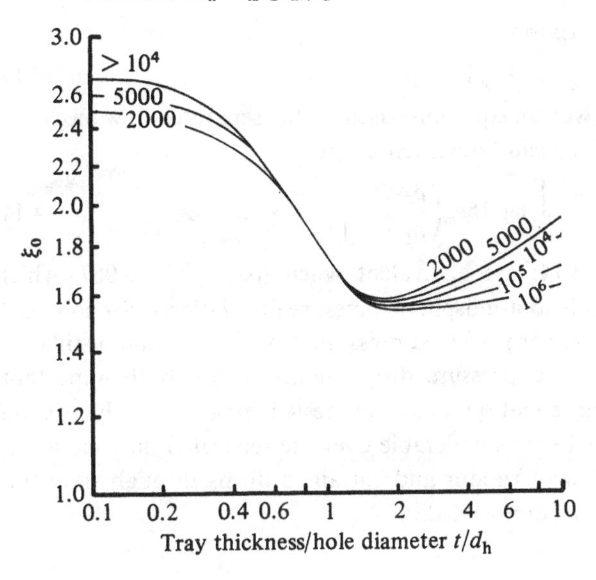

Fig. 4.3. Correlation of Smith & Van Winkle (1958) for orifice coefficient K_w. Parameter Re_h.

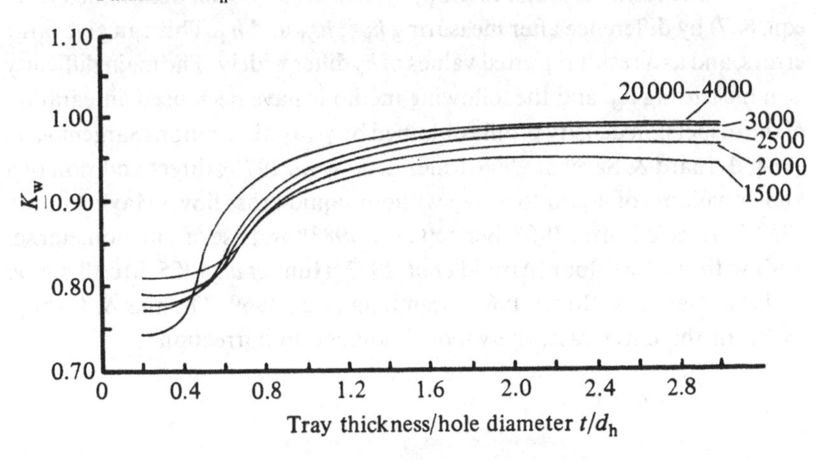

4.2.2 *Effect of entrainment on dry tray pressure drop*

The presence of entrained drops increases the dry tray pressure drop. Based on the homogeneous model of two-phase flow (Wallis 1969) the following equation can be derived:

$$h_{\mathrm{DT}} = (h_{\mathrm{DT}})_{e_{\mathrm{m}}=0}\left(1+\frac{e_{\mathrm{m}}\rho_{\mathrm{G}}}{\rho_{\mathrm{L}}}\right)(1+e_{\mathrm{m}}) \tag{4.12}$$

where

$$e_{\mathrm{m}} = \frac{\text{entrained liquid flow rate kg s}^{-1}}{\text{vapour flow rate kg s}^{-1}}$$

For all practical purposes,

$$h_{\mathrm{DT}} = (h_{\mathrm{DT}})_{e_{\mathrm{m}}=0}(1+e_{\mathrm{m}}) \tag{4.13}$$

Fair (1963) has derived an equation based on the separated flow model for two-phase flow which can be written as

$$h_{\mathrm{DT}} = (h_{\mathrm{DT}})_{e_{\mathrm{m}}=0}\left[1+15e_{\mathrm{m}}\left(\frac{\rho_{\mathrm{G}}}{\rho_{\mathrm{L}}}\right)^{0.5}\right] \tag{4.14}$$

Eqns. (4.13) and (4.14) are equivalent when $(\rho_{\mathrm{G}}/\rho_{\mathrm{L}})^{0.5}=0.067$, which typically corresponds to atmospheric pressure distillation – Section 1.3.1. Eqn. (4.13) gives a higher predicted pressure drop for vacuum distillation where entrainment and pressure drop prediction are both important. Apparently, neither equation has been tested experimentally. As the homogeneous model seems preferable over the separated flow model for representing the flow of vapour and entrained drops through sieve tray holes, eqn. (4.13) is recommended.

4.3 Residual pressure drop – sieve trays

The residual pressure drop, h_{R}, has usually been determined from eqn. (4.7) by difference after measuring h_{WT}, h_{DT} and h_{cl}. This can give large errors, and as a result reported values of h_{R} differ widely. The main difficulty is in measuring h_{cl}, and the following methods have been used: integration of the dispersion density profile obtained by γ-ray absorption (Sargent *et al.* 1964, Bernard & Sargent 1966, Pinczewski *et al.* 1975); direct addition of a known volume of liquid to a tray without liquid cross flow (Mayfield *et al.* 1952, Davies & Porter 1965, Bennett *et al.* 1983); and use of a manometer set flush with the tray floor (Arnold *et al.* 1952, Hunt *et al.* 1955, McAllister *et al.* 1958, Harris & Roper 1962, Brambilla *et al.* 1969, Thomas & Ogboja 1978). In the latter case, the vapour momentum correction

$$\frac{u_{\mathrm{s}}\rho_{\mathrm{G}}}{\rho_{\mathrm{L}}g}(u_{\mathrm{h}}-u_{\mathrm{s}})$$

usually has to be subtracted from the reported residual pressure drops to give h_R.

A major influence on h_R is the ratio t/d_h as shown in Fig. 4.4. Other workers have found $h_R \approx 10$ mm H_2O when $t/d_h = 1.0$ (Davies & Porter 1965, Hunt *et al.* 1955). It has also been found that h_R depends on the flow regime (Payne & Prince 1975, Pinczewski *et al.* 1973). A further complication is that h_R can apparently depend on as yet undetermined effects of column geometry (Thomas & Ogboja 1978). There is ample evidence that h_R falls with a reduction in surface tension (Mayfield *et al.* 1952, Hunt *et al.* 1955, Bennett *et al.* 1983). Residual pressure drop has traditionally been interpreted as the excess pressure required to overcome surface tension when bubbles are formed at the orifice (Van Winkle 1967, Fair 1963) so that

$$h_R = \frac{4\sigma}{d_h \rho_L g} \tag{4.15}$$

An alternative correlation proposed by Bennett *et al.* (1983) is

$$h_R = \frac{6}{b\rho_L}\left(\frac{\sigma}{g}\right)^{0.67}\left(\frac{\rho_L - \rho_G}{d_h}\right)^{0.33} \tag{4.16}$$

Fig. 4.4. Effect of tray thickness to hole diameter ratio on residual pressure drop. ● Experimental results of Brambilla *et al.* (1969); — average results of McAllister *et al.* (1958).

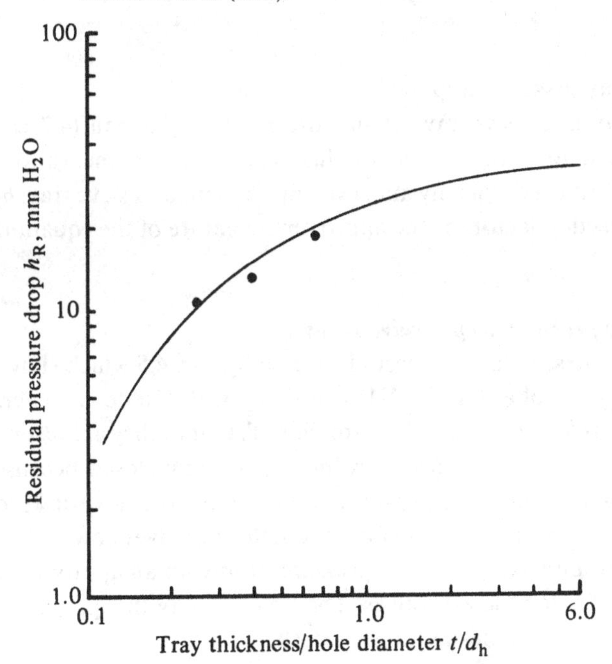

where b must be evaluated experimentally ($b = 1.27$ for air–water). The derivation leading to eqn. (4.7) indicates that eqns. (4.15) and (4.16) are oversimplified. Cervenka & Kolar (1974) have interpreted h_R as arising from partial blockage of the holes by liquid. Surface tension might be important here and Cervenka's correlation for the effective hole area does indeed involve surface tension.

A similar approach was proposed by Davy and Haselden (1975). They combined h_R with h_{DT} in eqn. (4.7) to give

$$h_{WT} = h'_{DT} + h_{cl} \tag{4.17}$$

where

$$h'_{DT} = \frac{\xi' \rho_G u_h^2}{2g\rho_L} \tag{4.18}$$

The modified orifice coefficient ξ' is related to the dry tray coefficient ξ by $\xi' = \xi/Z^2$, where

$$Z = 1 - \exp\{-2.65\, u_h [\rho_G/(\rho_L - \rho_G)]^{0.5}\} \tag{4.19}$$

Note that eqn. (4.19) attaches no significance to the surface tension or to the ratio t/d_h.

As there is no clear-cut 'best' way of predicting h_R it is recommended that Fig. 4.4 be used for design purposes. Non-aqueous systems will generally have a lower value of h_R than given by Fig. 4.4 so its use should err on the safe side.

4.4 Valve tray pressure drop

The analysis for sieve tray pressure drop which led to eqn. (4.7) can equally be applied to a valve tray. In this case, however, the various assumptions are less easy to justify and, even more than for a sieve tray, h_R represents a correction factor for the approximate nature of the equation.

4.4.1 *Dry tray pressure drop – valve trays*

Dry tray pressure drop is best illustrated by Fig. 4.5 which shows the dry pressure drop of a (Glitsch V1) valve tray with the downcomers blocked. At low gas flow rate the valves are 'closed', that is, they are resting on the tray deck. In fact, the valves are not completely closed because normal practice is to include tabs or protuberances under the valve disks to avoid the tendency of surface tension forces to hold the valves down when they start to open. Eqn. (4.8) gives the pressure drop with an appropriate orifice coefficient ξ_{VC} for 'closed' valves. The hole velocity in eqn. (4.8) is based on the area of the holes in the tray deck.

On further increasing the gas flow rate, the pressure under the valves is sufficient to cause them to lift. A simple force balance on a valve then gives

$$h_{DT} = \frac{\xi_{VD} u_h^2 \rho_G}{2g\rho_L} + \frac{KM_V}{\rho_L A_V} \tag{4.20}$$

where M_V is the mass of the valve and A_V is the disk area. The first term in eqn. (4.20) represents the pressure drop through the valve tray deck to the underside of the valve and involves a discharge coefficient for the deck. The second term gives the pressure drop to support the valve disk. The coefficient K accounts for radial pressure variations under the disk and various eddy losses. The gas velocity when the valves just start to open is the closed balance point (CBP).

Further increase in the gas flow rate causes additional valves to open to accommodate the extra gas flow while the pressure drop remains approximately constant and given by eqn. (4.20). Individual valves then open and close randomly over the tray with the proportion of valves which are open at any one time being fixed by the gas load. Any particular valve is either open or closed since, as shown by Wood (1961), a valve in an intermediate half-open position is unstable.

Eventually, as the gas load is increased even more, all valves remain open and the pressure drop is again given by eqn. (4.8) with an orifice coefficient

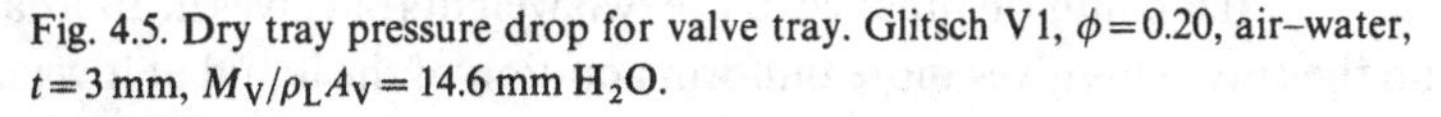

Fig. 4.5. Dry tray pressure drop for valve tray. Glitsch V1, $\phi = 0.20$, air–water, $t = 3$ mm, $M_V/\rho_L A_V = 14.6$ mm H_2O.

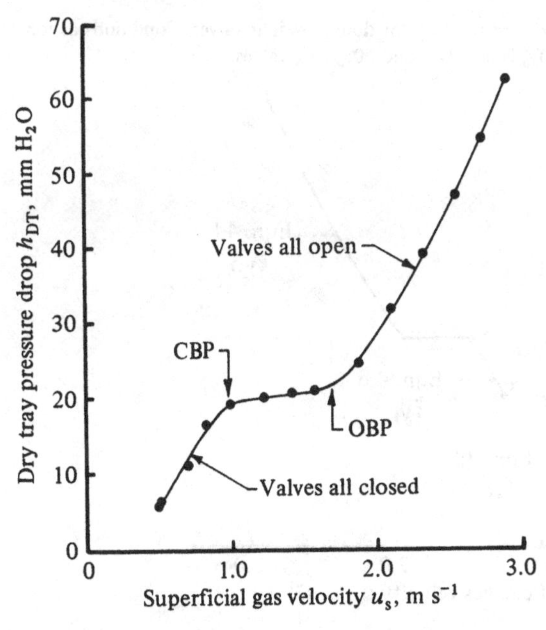

ξ_{VO} for open valves. The lowest gas velocity for which all the valves are open is the open balance point (OBP). Only the second term involving the valve mass is usually significant in eqn. (4.20) and, following Bolles (1976*a*), the superficial gas velocities at the balance points are given by equating eqns. (4.8) and (4.20) with $\xi_{VD}=0$ so that at the CBP

$$(u_s)_{CBP}=\phi\left[\frac{2gKM_V}{\rho_G\xi_{VC}A_V}\right]^{0.5} \tag{4.21}$$

and at the OBP

$$(u_s)_{OBP}=\phi\left[\frac{2gKM_V}{\rho_G\xi_{VO}A_V}\right]^{0.5} \tag{4.22}$$

where ϕ is the fractional open area of the tray deck.

The fraction of valves which are open v_o is given by

$$v_o=\frac{u_s-(u_s)_{CBP}}{(u_s)_{OBP}-(u_s)_{CBP}} \quad (0\leqslant v_o\leqslant 1) \tag{4.23}$$

The procedure for calculating the dry tray pressure drop is to first determine $(u_s)_{CBP}$ and $(u_s)_{OBP}$ from eqns. (4.21) and (4.22) and then v_o from eqn. (4.23). If $0<v_o<1$, eqn. (4.20) is used for h_{DT}. Otherwise eqn. (4.8) is used with $\xi=\xi_{VC}$ when $v_o=0$ and $\xi=\xi_{VO}$ when $v_o=1$.

4.4.1.1 *Double weight valves*

It is common practice to use two weights of valves in alternate rows on the tray. This gives more uniform aeration of the liquid at low gas loads

Fig. 4.6. Dry tray pressure drop for double weight valves. Conditions as for Fig. 4.5 except 50% 20 g valves and 50% 37 g valves.

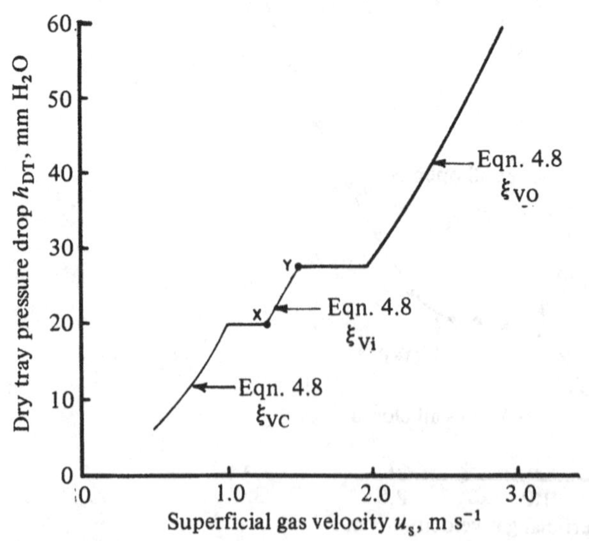

Table 4.1. *Dry tray pressure drop coefficients for valve trays*

Type	ξ_{VC}	ξ_{VD}	K	ξ_{VO}	Reference
Standard Glitsch, Koch and Nutter	$1.9\left(\frac{A_h}{A_{sc}}\right)^2$		1.2–1.4	$\frac{7.9}{t^{0.5}}-\frac{8.3}{t^{0.5}}$	Bolles (1976*a*)
Venturi Glitsch, Koch and Nutter	$1.9\left(\frac{A_h}{A_{sc}}\right)^2$		1.2–1.4	$\frac{4.0}{t^{0.5}}-\frac{4.8}{t^{0.5}}$	Bolles (1976*a*)
Glitsch V1		1.1	1.35	5.63–3.1	Glitsch, Inc. (1974)
Glitsch V4		0.54	1.35	2.7	Glitsch, Inc. (1974)
Glitsch V1		1.95[a]	0.77[a]	4.4[a]	Piqueur & Verhoeye (1976)
Nutter		0.99	0.82	3.0	Brambilla & Nencetti (1973)
Various		1.27	0.85	4.2	Brambilla *et al.* (1969)
Special				$200/l^2$	Dytnerskii *et al.* (1965)

[a] Only eight valves/tray
l = maximum valve lift (mm) $1<l<8$
A_h = area of holes in tray deck
A_{sc} = total vertical curtain area through which vapour passes as it leaves less area obstructed by valve legs or tabs
t = deck thickness (mm)

and also by judicious selection of valve weights can reduce tray pressure drop while still avoiding weeping (Bolles 1976*a*). Fig. 4.6 shows the dry tray pressure drop for a double weight valve tray and the four balance points involved. For n_1 light valves and n_2 heavy valves, the intermediate orifice coefficient used in eqn. (4.8) for pressure drops between points X and Y on Fig. 4.6 is given by

$$\xi_{\mathrm{Vi}} = (n_1 + n_2)^2 \left[\frac{n_1}{\xi_{\mathrm{VO}}^{0.5}} + \frac{n_2}{\xi_{\mathrm{VC}}^{0.5}} \right]^{-2} \tag{4.24}$$

4.4.1.2 *Orifice coefficients for valve trays*

A summary of recommended orifice coefficients for valve trays for use with eqn. (4.8) is shown in Table 4.1. Also shown are values of K for use in eqns. (4.20)–(4.22).

4.4.2 ***Residual pressure drop – valve trays***

Considerable confusion exists regarding the residual pressure drop for valve trays, and negative values have been reported by Brambilla *et al.* (1969). The difficulty lies in making accurate measurements of the clear liquid height on valve trays for use in eqn. (4.7). At present, residual pressure drop is best ignored when estimating valve tray pressure drop.

5

Maximum capacity

5.1 Flooding

5.1.1 *Introduction*

Flooding occurs when a particular combination of vapour and liquid flow rates exceeds the capacity of the column. A flooded tray acts as a restriction to the liquid flowing down the column. Thus, a common symptom of flooding is a build-up of liquid and an excessive pressure drop over the column section above the flooded tray.

There are two distinct types of flooding: downcomer flooding and jet flooding, and these correspond to the way in which flooding starts. Downcomer flooding usually occurs at high flow parameters when a combination of an excessive liquid flow rate and a tray spacing which is too small causes the froth in the downcomer to back up over the weir of the tray above. This represents the maximum capacity of the downcomer. Jet flooding generally occurs at lower flow parameters and is initiated by excessive liquid entrainment. As shown in Section 4.2.2, liquid entrainment increases the tray pressure drop. This, together with the extra liquid load caused by recycled liquid (which has to leave the tray via the downcomer), contribute to increased backup of froth in the downcomer.

Note that the final stage in both types of flooding is the same – backup of froth in the downcomer to the weir of the tray above.

We are concerned here with the maximum expected capacity of the trays. Premature flooding can occur below the maximum expected capacity because of such things as sloppy tray installation, poor design of feed and side draw trays, physical blockage, etc. Some 20 possible causes of premature flooding have been documented (AIChE 1986).

5.1.2 *Flooding correlations*

It is difficult to predict the onset of flooding using a fundamental approach because correlations for liquid entrainment and downcomer froth density are insufficiently accurate. Consequently, the approach taken is to size the tray bubbling area using an empirical jet flooding correlation and to size the downcomers using various rules based on experience. Checks are subsequently made for excessive entrainment or downcomer backup.

Fair's flooding correlation (Fair 1961) for sieve trays having $d_h \leqslant 6.4$ mm is shown in Fig. 5.1. Capacity factor CF' is defined as

$$CF' = u_s'\left(\frac{\rho_G}{\rho_L - \rho_G}\right)^{0.5} \tag{5.1}$$

where

$$u_s' = \frac{Q_G}{A_n}$$

and A_n is the net area for liquid disengagement above the tray, i.e. for single-pass trays, $A_n = (\pi D^2/4) - A_d = A_b + A_d$.

Fair's recommendations for correcting the capacity factor from Fig. 5.1, CF'', for surface tension and fractional open area of the tray can be expressed as

$$CF' = CF''\left(\frac{\sigma}{20 \times 10^{-3}}\right)^{0.2}\left(\frac{\phi}{0.1}\right)^{0.44} \tag{5.2}$$

Fig. 5.1. Fair's correlation for sieve tray flooding (Fair 1961). $d_h \leqslant 6.4$ mm, $h_w < 0.15\ T_s$.

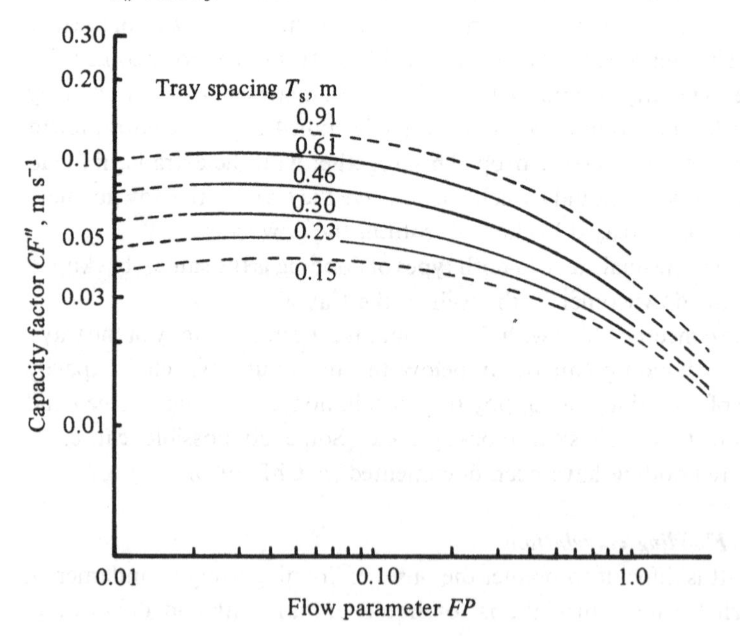

The correction for fractional tray open area, ϕ, is only required when $\phi < 0.1$. It reflects an increased tendency for vapour jet penetration through the froth at high hole vapour velocities. Values of CF'' from Fig. 5.1 can be approximately represented by the following equation (Treybal 1968):

$$CF'' = (0.0744\,T_s + 0.0117)(\log_{10} FP^{-1}) + 0.0304\,T_s + 0.0153 \tag{5.3}$$

When $FP < 0.1$, FP is set equal to 0.1 in eqn. (5.3).

In contrast to Fair's correlation, most recent jet flooding correlations involve the capacity factor based on the tray bubbling area, CF, and the liquid weir load, Q_L/W. For commercial reasons these correlations are often wrapped up into a different form.

The Glitsch correlation (Glitsch, Inc., 1974) for a single-pass valve tray can be rearranged to give

$$CF = CF_0 \cdot FF \cdot SF - \frac{Q_L}{W} \tag{5.4}$$

where

$$CF = \frac{Q_G}{A_b}\left(\frac{\rho_G}{\rho_L - \rho_G}\right)^{0.5} \tag{5.5}$$

Eqn. (5.4) is based on an assumed liquid flow path length of 0.75 D. The capacity factor at zero liquid load, CF_0, is a function of tray spacing and vapour density and is given in Fig. 5.2. Two safety factors are used. The fractional approach to flooding, FF, is typically taken as 0.8 and the system

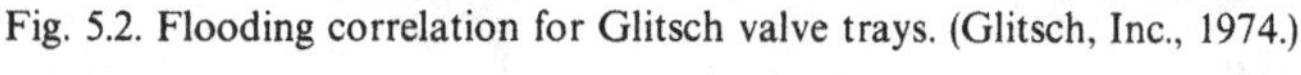
Fig. 5.2. Flooding correlation for Glitsch valve trays. (Glitsch, Inc., 1974.)

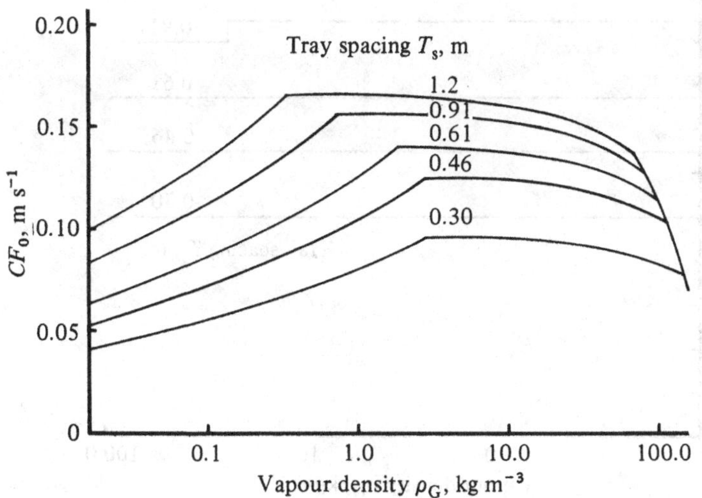

factor, SF, lowers the design value of the capacity factor when the system is known to foam. Typical values of SF have been given in Table 2.2.

The Koch correlation for valve trays (Koch Engineering Co. 1982) may similarly be rearranged to give

$$CF = CF_0 \cdot FF \cdot SF - 1.96\frac{Q_L}{W} \quad (5.6)$$

The factor CF_0 may be obtained from Fig. 5.3.

A slightly different type of flooding correlation is used for Nutter valve trays (Nutter Engineering Co. 1976) and is

$$CF = CF_0 \cdot FF \cdot SF - \frac{Q_L}{A_b} \quad (5.7)$$

Eqn. (5.7) shows a different dependence on liquid load then eqns. (5.4) and (5.6), but it does have the merit of dimensional homogeneity. The capacity factor at zero liquid load, CF_0, is a function of both σ and ρ_G as well as T_s – Fig. 5.4. Note that for a new column design both eqns. (5.4) and (5.6) require an iterative calculation because W is initially unknown, whereas Fair's correlation and eqn. (5.7) can be used directly.

Fig. 5.5 shows how these flooding correlations compare with sieve tray flooding data released by Fractionation Research, Inc. (FRI) (Sakata & Yanagi 1979). All the valve tray correlations give similar results, which seems to indicate that minor variations in valve design are unimportant in determining flooding.

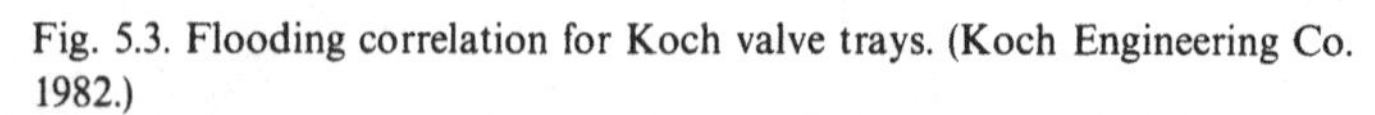

Fig. 5.3. Flooding correlation for Koch valve trays. (Koch Engineering Co. 1982.)

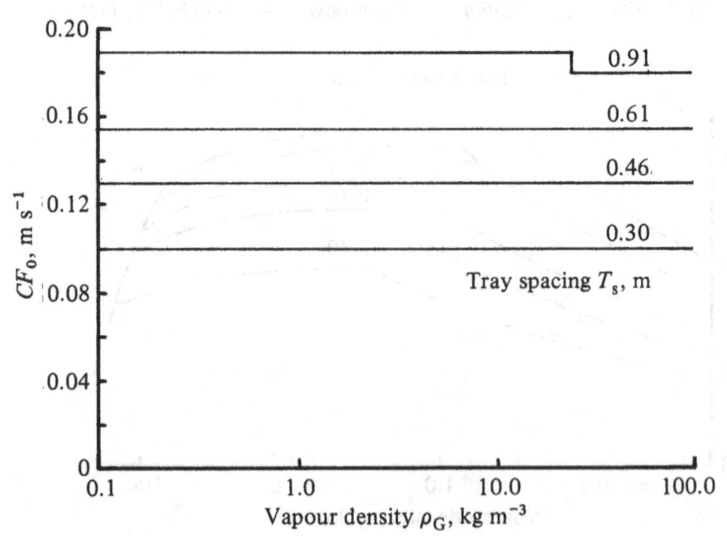

The FRI data indicates a maximum vapour capacity at $Q_L/W = 0.01\,m^3\,m^{-1}\,s^{-1}$. This corresponds to a minimum in the measured entrainment rate (Porter & Jenkins 1979) and is related to the transition between the spray and froth regimes. Low liquid loads give low clear liquid heights, and at high vapour velocities this results in spray regime operation.

Fig. 5.4. Flooding correlation for Nutter valve trays. (Nutter Engineering Co. 1976.)

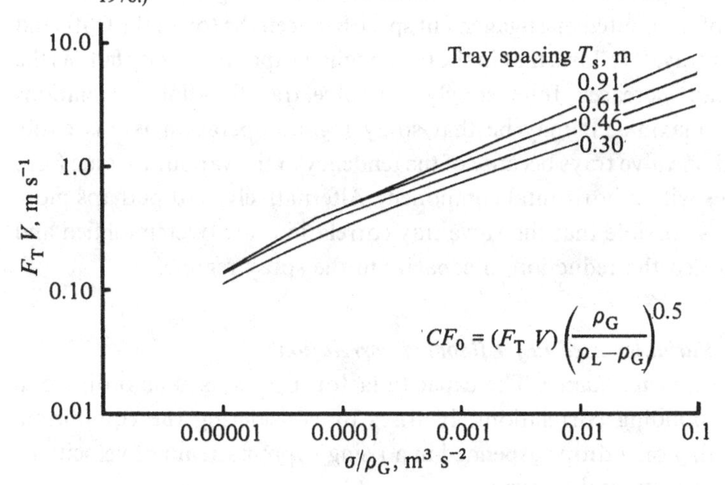

Fig. 5.5. Comparison of jet flooding correlations with FRI data. Cyclohexane–*n* heptane at 165 kPa. Experimental points denoted by ●.

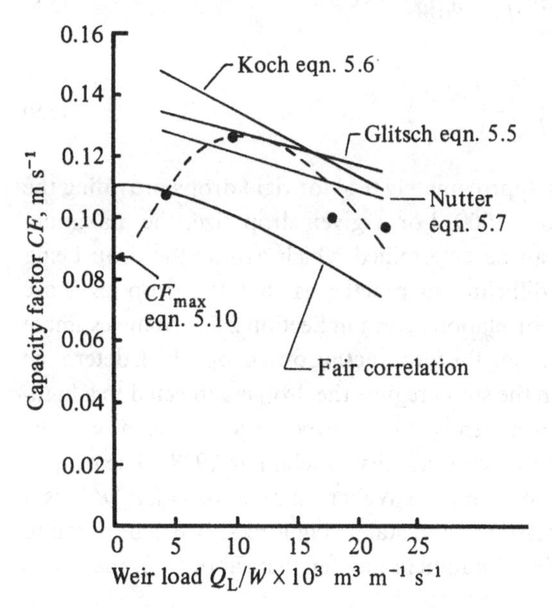

Since entrainment is high in the spray regime, vapour capacity is limited. As the liquid load is increased, while remaining in the spray regime, clear liquid height also increases. Because the vapour now passes through more liquid, it is slowed down by transfer of momentum to the liquid. The lower velocity vapour jets, which then leave the pool of liquid on the tray, generate less entrainment and vapour capacity increases. In the froth regime, to the right of the maximum on Fig. 5.5, an increase of liquid load causes an increase in the froth height on the tray. Entrainment then progressively increases because of the limited disengagement space between the top of the froth and the upper tray. It follows that in the froth regime vapour capacity falls as the liquid load increases. Interestingly the valve tray flooding correlations show no maxima. It may be that spray regime operation is less easily obtained on valve trays because of the tendency of the vapour to issue from the valves with a horizontal component. Alternatively, and perhaps more likely, it is possible that the valve tray correlations are oversimplified and have masked the reduction in capacity in the spray regime.

5.1.3 *Variables used in jet flooding correlations*

Capacity factor. The capacity factor, CF, is used as ordinate in most jet flooding correlations. It arises by considering the equilibrium forces acting on a drop suspended in a rising vapour stream of velocity u_s, i.e. drag, gravity and buoyancy.

$$C_0 \frac{\pi d_p^2}{4} \frac{\rho_G}{2} u_s^2 = \frac{\pi d_p^3}{6} (\rho_L - \rho_G) g \tag{5.8}$$

so that

$$CF = u_s \left(\frac{\rho_G}{\rho_L - \rho_G} \right)^{0.5} = \left(\frac{4 d_p g}{3 C_0} \right)^{0.5} \tag{5.9}$$

The drag coefficient C_0 is approximately 0.44 for rigid drops providing the Reynolds number is above 1000. For a given drop size, the maximum permissible value of u_s can be determined which avoids the drop being entrained upward. The difficulty in practice is that the drop sizes are unknown, in spite of the correlations given in Section 2.7.3. Some estimate can be obtained by considering the forces acting on a drop which determine its size (Manning 1964). In the spray regime the drop is subjected to a high-velocity vapour jet which tends to destroy it and which can be approximated by the hole vapour velocity (Stichlmair 1978). The drop is stabilised by surface tension, and the Weber number $We = d_p \rho_G u_h^2 / \sigma$ is a measure of the ratio of destructive to stabilising forces. It has been found (Wallis 1969) that for Weber numbers greater than about 12, the drops become unstable for low-viscosity liquids, so that an estimate of the maximum drop size is

$$d_{\text{pmax}} = \frac{12\sigma}{\rho_G}\left(\frac{\phi}{u_s}\right)^2$$

Combining this with eqn. (5.9)

$$CF_{\text{max}} = 2.5\left(\frac{\sigma g}{\rho_L - \rho_G}\right)^{0.25} \phi^{0.5} \tag{5.10}$$

Eqn. (5.10) and Fair's empirical corrections for σ and ϕ, eqn. (5.2), are in good agreement. This is remarkable since eqn. (5.10) can be expected to apply only in the fully developed spray regime at very low liquid loads. Only in this case is all the liquid on the tray in contact with vapour having a velocity equal to the hole vapour velocity as assumed in its derivation. At higher liquid loads, the vapour jets quickly slow down from the hole velocity as a result of momentum transfer, so that when they issue from the liquid they have a velocity somewhere between the hole vapour velocity and the superficial vapour velocity. Nevertheless, eqn. (5.10) can be used to predict CF_{max} at zero liquid load. The predicted value is included on Fig. 5.5 and is consistent with the FRI data extrapolated to zero liquid load.

Liquid weir load or flow parameter? There is less agreement about the appropriate abscissa to use in jet flooding correlations – Fair's correlation uses FP and most others use Q_L/W. The use of FP arises naturally in situations where the two phases flow countercurrently to each other, as in packed columns, but at first sight there seems to be no fundamental reason why it should also be used for the cross flow situation on trays. However, in distillation, changes in FP are closely matched by changes in the liquid weir load as shown below.

At total reflux, $M_G = M_L$ and $Q_G\rho_G = Q_L\rho_L$. The various jet flooding correlations, Fig. 5.5, indicate that for a fixed tray spacing, the bubbling area required strongly depends on $Q_G(\rho_G/\rho_L)^{0.5}$. As an approximation, the smaller effect of liquid load on the bubbling area required is neglected. It follows that

$$A_b \propto Q_G(\rho_G/\rho_L)^{0.5} = Q_L(\rho_L/\rho_G)^{0.5} \tag{5.11}$$

where $\propto$ means 'is proportional to'. Dividing by W,

$$\frac{A_b}{W} \propto \frac{Q_L}{W}\left(\frac{\rho_L}{\rho_G}\right)^{0.5} \tag{5.12}$$

For one or two passes, we have, approximately,

$$A_b \propto WD \tag{5.13}$$

so that

$$\frac{Q_L}{W} \propto D\left(\frac{\rho_G}{\rho_L}\right)^{0.5} \tag{5.14}$$

and

$$\frac{Q_L}{W} \propto D \cdot FP \tag{5.15}$$

at total reflux. Although approximate, eqn. (5.15) indicates that the liquid weir load increases as both column diameter and pressure increase. This explains why large-diameter high-pressure distillation requires multipass trays to avoid excessive liquid weir loads – Section 1.3.3. Eqn. (5.15) also shows that an increase in FP is associated with an increase in liquid weir load. It follows that FP is a satisfactory alternative to Q_L/W in flooding correlations, at least for distillation. The above arguments, of course, break down for absorption and stripping where there is only a weak relationship between M_G and M_L.

5.2 Liquid entrainment

It is important to estimate the rate of entrainment, both as a check on the approach to flooding and also to estimate the reduction in tray efficiency caused by entrainment.

5.2.1 *Mechanism of entrainment*

Two mechanisms of entrained drop generation have been identified. Drops are created by gas rush and film rupture as a bubble bursts (Newitt *et al.* 1954) and, at the other extreme, by the shearing action of a vapour jet which produces liquid sheets, ligaments and eventually drops (Nielsen *et al.* 1965). The actual mechanism in practice is some combination of these extreme cases depending on the flow regime. Numerous authors have shown that different entrainment correlations are required in the spray and froth regimes (Shakhov *et al.* 1964, Dytnerskii & Andreev 1966, Banerjee *et al.* 1969*b*, Lockett *et al.* 1976, Kister *et al.* 1981*a*).

In the spray regime, those factors which tend to push the tray further into the spray regime also tend to increase entrainment. As a result, entrainment increases with increased hole vapour velocity and hole diameter and decreases with increased liquid weir load and increased fractional perforated area. Similar behaviour is found in the mixed-froth regime where entrainment generation by vapour jets is still the predominant mechanism. In the froth regime proper and in the emulsion regime, entrainment primarily depends on the approach of the upper surface of the froth to the tray above. Those factors which increase froth height also increase entrainment, and so the latter increases with vapour velocity, liquid load and weir height. In all regimes, a reduction of tray spacing increases entrainment. Surface tension is also important, with a high surface tension tending to stabilise drops and so lower entrainment. In the froth and emulsion regimes foaming raises the dispersion height and increases entrainment (Elenkov & Vulchev 1968, Garvin & Norton 1968, Assenov *et al.* 1969).

5.2.2 *Measurement of entrainment*

The following are the main techniques which have been used to measure entrainment. In general, each technique gives a different estimate of the entrainment rate, so it is convenient to think in terms of four 'types' of entrainment.

Free entrainment – measured by sampling drops between the top of the froth/spray and the tray above (Akselrod & Yusova 1957, Aiba & Yamada 1959, Cheng & Teller 1961, Banerjee *et al.* 1969*a*, *b*).

Dry tray entrainment – measured by installing a special 'dry' tray above the operating tray and collecting the liquid which accumulates on it (Hunt *et al.* 1955, Friend *et al.* 1960, Bain & Van Winkle 1961, Lemieux & Scotti 1969, Molokanov *et al.* 1969, Sakata & Yanagi 1979).

Wet tray entrainment – measured by introducing a non-volatile tracer into the liquid on a tray and by sampling the liquid leaving the tray above for tracer (Strang 1934, Holbrook & Baker 1934, Kamei *et al.* 1951, Jones & Pyle 1955, Molokanov *et al.* 1969, Lockett *et al.* 1976).

Efficiency entrainment – estimated from the reduction in tray efficiency caused by entrainment (Aleksandrov & Shoblo 1962, Teller *et al.* 1963, Kastanek & Standart 1967).

Measurement of wet tray entrainment is preferred because it is a direct method which can be used under actual operating conditions either on the plant or in the laboratory. It has a high accuracy providing the tracer does not change the surface tension. So far its reported uses have been confined to small completely mixed trays. Measurement of dry tray entrainment, although probably the most common technique, suffers because it has to be assumed that no re-entrainment occurs from the dry tray. Also the amount of liquid present on the dry tray probably influences the amount of entrainment it collects from the tray below (Piqueur & Verhoeye 1976). Free entrainment is relatively easy to measure under laboratory conditions, by using a coated slide for example, but it is usually larger than the entrainment captured by the upper tray, although not always, as shown by Calcaterra *et al.* (1968). Efficiency entrainment is open to very large errors because of the difficulty of measuring the efficiency with sufficient accuracy and then of relating it to entrainment. Consequently, this method is unsuitable for use in industrial columns.

5.2.3 *Prediction of entrainment*

Early correlations took no account of the flow regime. Of this type of correlation, that of Fair & Matthews (1958) and of Eduljee (1958*b*) are the most comprehensive for bubble cap trays. For sieve trays, Fair's correlation (Fair 1961) is popular and easy to use – Fig. 5.6. The approach to flooding can be obtained from Fig. 5.1. At low values of *FP*, corresponding to the spray regime, entrainment rises rapidly.

More recent work for sieve trays has concentrated on developing separate correlations for the spray and froth regimes. Most are elaborations of the simple correlation due to Hunt *et al.* (1955):

$$\frac{E_m}{M_G} = 7.75 \times 10^{-5} \left(\frac{73 \times 10^{-3}}{\sigma} \right) \left(\frac{u_s}{T_s - h_f} \right)^{3.2} \tag{5.16}$$

In deriving eqn. (5.16), an assumed value of $\alpha = 0.4$ was used to estimate h_f from measured clear liquid heights. To be consistent, the same assumption should also be made when using eqn. (5.16) to estimate entrainment. Nearly all the data used to develop eqn. (5.16) were obtained in the froth regime and consequently there was very little influence of d_h or ϕ.

Fell has recently developed correlations which are specific to a particular

Fig. 5.6. Entrainment correlation for sieve trays due to Fair (1961). $d_h \leqslant 6.4$ mm.

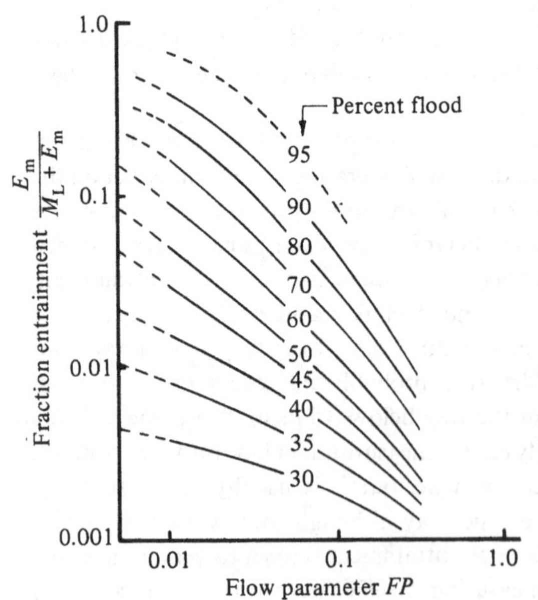

regime (Kister *et al.* 1981*a*, *b*, Fell & Pinczewski 1982) and they can be represented by

$$\frac{E_m}{M_G}=K_1\left[\frac{u_s}{(Q_L/W)^{0.25}}\frac{1}{(1+K_2h_w)}\frac{h_{cl}}{(d_hT_s)^{0.5}}\right]^{4.68}\left(\frac{\mu_G}{\sigma}\right)^{1.17} \tag{5.17}$$

In the spray regime, $K_2=0$ with $K_1=13.1$ for air–water and $K_1=30.6$ for hydrocarbons. In the mixed-froth regime, $K_2=2.62$ and for air–water $K_1=20.5$.

The clear liquid height, h_{cl}, is obtained from a correlation which is strictly only applicable at the froth–spray transition (Jeronimo & Sawistowski 1973)

$$h_{cl}=\frac{0.694(gd_h^5/\sigma(\rho_L-\rho_G)^2)^{0.167}\phi^{-0.791}}{1+1.04\times10^{-4}(Q_L/W)^{-0.59}\phi^{-1.79}} \tag{5.18}$$

(There is an error in the equation quoted by Jeronimo for air–water.)

Although eqn. (5.17) is of a similar form to eqn. (5.16), it also involves d_h and ϕ (through eqn. (5.18)) which are found to be of importance in the spray and mixed-froth regimes. Eqn. (5.17) does an adequate job of correlating the available air–water data but it has not been extensively tested for hydrocarbon systems.

Zuiderweg (1982) developed an entrainment correlation for spray regime hydrocarbon systems based on FRI data:

$$\frac{E_m}{M_G}=1\times10^{-8}\phi^{-2}\left(\frac{h_{cl}}{T_s}\right)^3\left(\frac{M_G}{M_L}\right)\left(\frac{\rho_L}{\rho_G}\right)^2\left\{1+265\left[\frac{u_s}{(gh_{cl})^{0.5}}\left(\frac{\rho_G}{\rho_L}\right)^{0.5}\right]^{1.7}\right\}^3 \tag{5.19}$$

where h_{cl} is determined from eqn. (3.26).

Eqn. (5.19) correlates the available FRI data remarkably well. However, it should be used with caution because it is based on eqn. (3.26) for h_{cl} in the spray regime and the influence of h_w on h_{cl} in this regime has not been confirmed. Surprisingly, neither d_h or σ are involved in Zuiderweg's correlation. Possibly the influence of σ is disguised by the inclusion of ρ_G/ρ_L, since, for the systems used by FRI, changes in physical properties can be correlated against each other – Section 1.3.1. Furthermore, d_h was held constant at 12.7 mm so eqn. (5.19) must be considered specific for this hole size. Note also that eqn. (5.19) is based on dry tray entrainment, whereas an estimate of the wet tray entrainment is required in practice.

Fig. 5.7 shows an entrainment correlation due to Stichlmair (1978) which includes data from sieve, valve and bubble cap trays. Two regimes are evident: the froth regime and the spray regime. In the froth regime, the disengaging space T_s-h_f is introduced. The correlation vividly indicates the rapid rise in entrainment in the spray regime although it fails to reflect any influence of hole diameter in this regime.

Entrainment measurements from valve trays have also been reported (Tasev & Elenkov 1970, Piqueur & Verhoeye 1976) and a number of Russian studies have provided data for grid trays (Val'dberg *et al.* 1969, Mikhailenko *et al.* 1975), for film trays (Chekov *et al.* 1975), and for various other special tray types (Aleksandrov & Shoblo 1960, Sum-Shik *et al.* 1963, Malafeev & Malyusov 1973).

At the present state of development, it is difficult to recommend any of these entrainment correlations as the universal best one to use under all circumstances. The prudent designer should obtain estimates from more than one of them. When entrainment appears to be more than 5% of the liquid flow rate, remedial steps should be taken at the design stage. The usual options are to increase the tray spacing or the bubbling area of the tray. In the froth regime, consideration should be given to reducing the weir height. In the spray regime reducing d_h or increasing ϕ both tend to reduce entrainment.

5.3 Downcomer hydraulics

5.3.1 *Downcomer backup*

Liquid is conveyed through the downcomer from a lower to a higher pressure and consequently liquid backs up in the downcomer to overcome the pressure difference. The primary requirement of a

Fig. 5.7. Entrainment correlation for trays due to Stichlmair (1978). $d_h \leqslant 10$ mm.

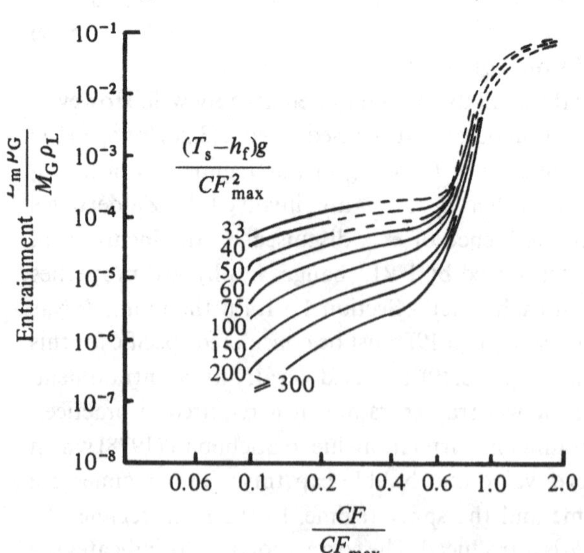

downcomer is that its height must be sufficient to accommodate this backup to avoid flooding.

From a pressure balance (Fig. 5.8), the downcomer backup h_{fd} is given by

$$h_{fd} = \frac{(h_{WT} + h_{cli} + h_{udc} - h_n)}{\bar{\alpha}_d} \tag{5.20}$$

At high pressures eqn. (5.20) can be elaborated to take vapour density into account. In spite of its simplicity, some of the terms in eqn. (5.20) are difficult to predict.

The wet tray pressure drop, h_{WT}, is the dominant term in eqn. (5.20) and fortunately it can be predicted with reasonable accuracy; see Chapters 3 and 4. The clear liquid height at liquid entry, h_{cli}, can be estimated in various ways:

(*a*) The simplest assumption is to take h_{cli} equal to the clear liquid height on the tray.

(*b*) If hydraulic gradient is significant, h_{cli} can be increased appropriately.

(*c*) As shown in Section 3.7.4, the liquid entry region is typically one of increased clear liquid height. The theory given there can be used to estimate h_{cli} at liquid entry ($h_{cli} = h_i$). Alternatively, Hofhuis (1980) has suggested that, for an initial unperforated region of 100 mm, $h_{cli} = 1.6\, h_{cl}$, where h_{cl} is the average value on the fully aerated part of the tray.

Fig. 5.8. Nomenclature for liquid backup in the downcomer.

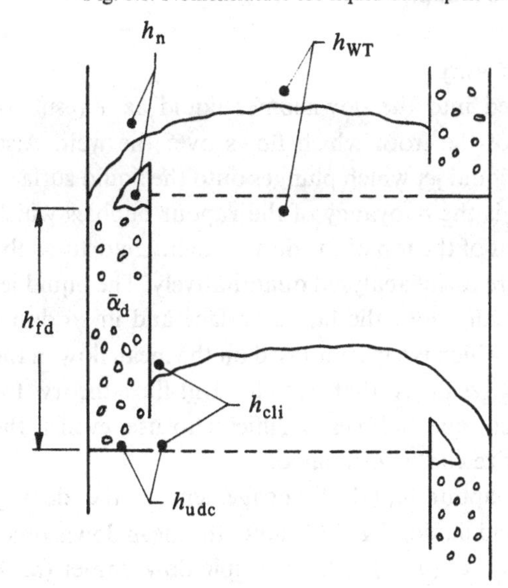

(*d*) When a hydraulic jump occurs at the liquid inlet, h_{cli} can be equated to the clearance under the downcomer. However, if the liquid is partly aerated as it passes under the downcomer, this will reduce h_{cli}.

The pressure drop under the downcomer h_{udc} is given by

$$h_{udc}=\frac{1}{2g}\left(\frac{Q_L}{Wh_1C_d}\right)^2 \tag{5.21}$$

Values for the discharge coefficient C_d have been given by Cicalese *et al.* (1947*a*, *b*). Aerated liquid flowing under the downcomer appears to have little effect on C_d (Lockett & Gharani 1979). Determination of C_d from experimental data is very dependent on the precise location of the downstream manometer (Hofhuis 1980). Some designers prefer to use a radius edge clearance under the downcomer to increase C_d. Values of C_d used by tray vendors are 0.54 (Glitsch, Inc. 1974), 0.56 (Koch Engineering Co. 1982), 0.60 (Nutter Engineering Co. 1976) for a standard sharp edge and 1.0 for a radius edge clearance. If an inlet weir or a recessed receiving pan is used, then h_{udc} is usually increased by about 20% over the value given by eqn. (5.21) with h_1 then being taken as the minimum clearance for flow.

A pressure increase across the nappe, h_n, is created when the throw of froth over the weir acts as a seal at the mouth of the downcomer. It is significant for narrow downcomers (Lockett & Gharani 1979) and for foaming systems (Thomas & Shah 1964). It is almost always neglected in calculating downcomer backup at the design stage since it acts conservatively and reduces backup.

5.3.2 *Downcomer froth density*

Vapour is entrained into the downcomer liquid as a result of ineffective disengagement of the froth which flows over the weir. Also vapour is entrained by the liquid jet which plunges onto the liquid surface; see Fig. 5.9. Opposing this is the buoyancy of the vapour bubbles which tend to rise upwards and out of the top of the downcomer. Unfortunately, none of these phenomena are easily analysed quantitatively. The liquid jet usually plunges asymmetrically onto the liquid surface and in so doing generates liquid circulation which is superimposed on the mean flow in the downward direction. The result is that simple drift-flux theory for countercurrent flow of liquid and bubbles is difficult to use, even if the appropriate mean bubble size can be estimated.

The driving force for vapour–liquid disengagement is the density difference between vapour and liquid. Fig. 5.10 shows the mean downcomer froth density $\bar{\alpha}_d$ as a function of $(\rho_L-\rho_G)$ for multiple downcomer (MD) trays (Union Carbide 1970). Low values of $\bar{\alpha}_d$ occur at high pressures corresponding to low values of $(\rho_L - \rho_G)$, and these are typically associated with light hydrocarbon distillation. An additional consideration which determines $\bar{\alpha}_d$ is the flow regime which exists on the tray. When the emulsion

Fig. 5.9. Liquid and vapour circulation in a downcomer. (Lockett & Gharani 1979.)

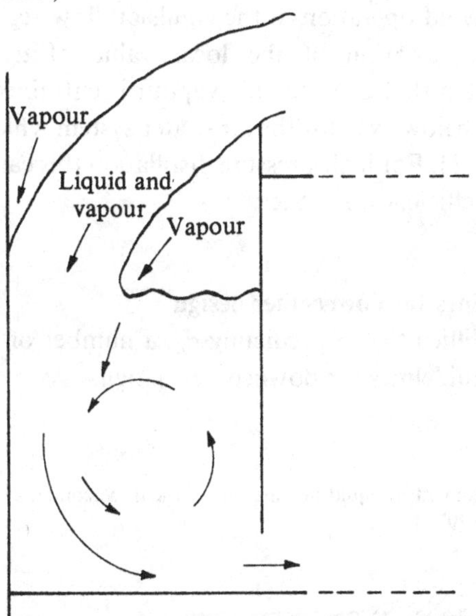

Fig. 5.10. Liquid volume fraction in downcomer for non-foaming systems. × determined from FRI data for conventional trays.

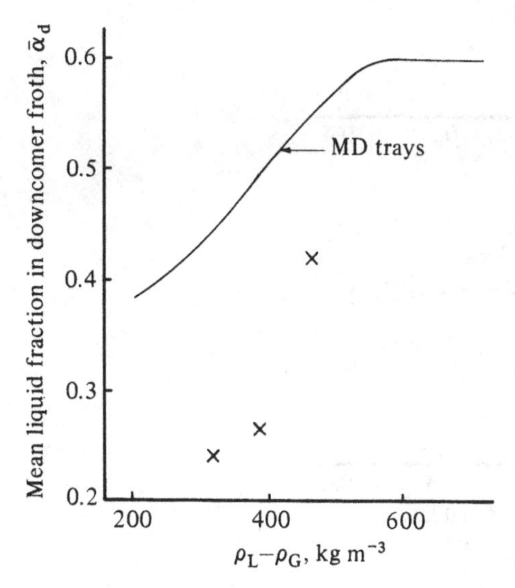

regime occurs, considerable amounts of vapour are entrained into the downcomers as small bubbles. Also shown on Fig. 5.10 are $\bar{\alpha}_d$ values determined by Hoek & Zuiderweg (1982) from FRI distillation data where the tray operated in the emulsion regime. Evidently, the combination of emulsion flow and high pressures gives $\bar{\alpha}_d$ values somewhat lower than those shown for the MD tray, and it is noteworthy that MD trays are specifically designed to avoid operation in the emulsion flow regime.

Measurements of the variation of the local value of α_d within downcomers have indicated that considerable vapour is entrained in the liquid flowing to the tray below even for the air–water system which has a high value of $\bar{\alpha}_d$; see Fig. 5.11. For high-pressure distillation this can have a significant effect on tray efficiency; see Section 9.12.

5.4 Empirical guidelines for downcomer design

Because of the difficulties of predicting $\bar{\alpha}_d$, a number of rules of thumb have evolved as guidelines for downcomer design.

Fig. 5.11. Gas entrainment in liquid flowing out of the downcomer, air–water. (Lockett & Gharani 1979.)

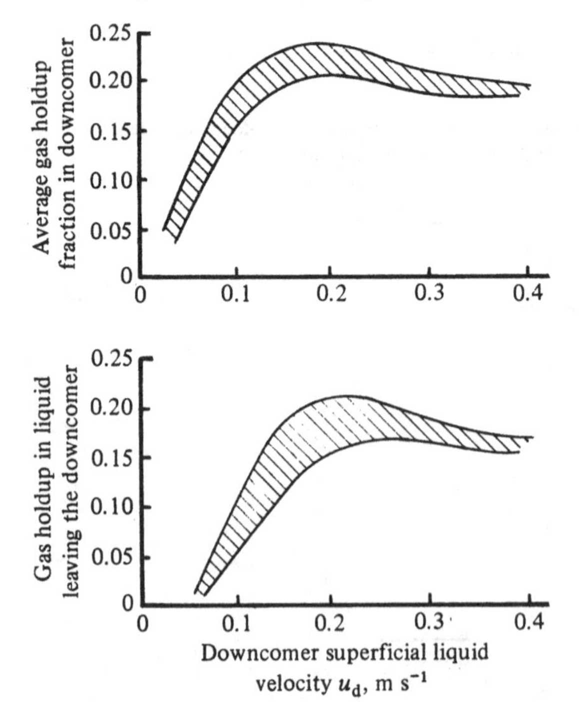

5.4.1 *Liquid velocity*

Downcomer area is usually fixed by specifying the superficial velocity of the liquid in the downcomer on a vapour-free basis u_d. A convenient equation due to Glitsch, Inc. (1974), is

$$u_d = \frac{Q_L}{A_d} = 0.0081(T_s)^{0.5}(\rho_L - \rho_G)^{0.5} \tag{5.22}$$

Other tray vendors express u_d graphically, and Fig. 5.12 indicates that there is poor agreement between them. By restricting u_d to the values shown in Fig. 5.12, it is found that, for non-foaming systems, $\bar{\alpha}_d = 0.5$ gives a safe conservative estimate for calculating downcomer backup using eqn. (5.20). For foaming systems, downcomers should be designed for lower values of u_d by multiplying the values from Fig. 5.12 by a system factor taken from Table 2.2. To determine downcomer backup, the value of $\bar{\alpha}_d$ appropriate to the foaming system should be used in eqn. (5.20). Unfortunately, there is very little published information on $\bar{\alpha}_d$ for foaming systems and this constitutes a significant part of the 'know how' of tray vendors.

5.4.2 *Liquid weir load*

High weir loads lead to large froth heights and the need to use excessive tray spacings. They also contribute to emulsion flow, hydraulic gradient problems, hydraulic jumps and choking and bridging of the downcomer mouth. To avoid excessive weir loads the number of passes is usually increased. The maximum weir load is a matter of judgement about

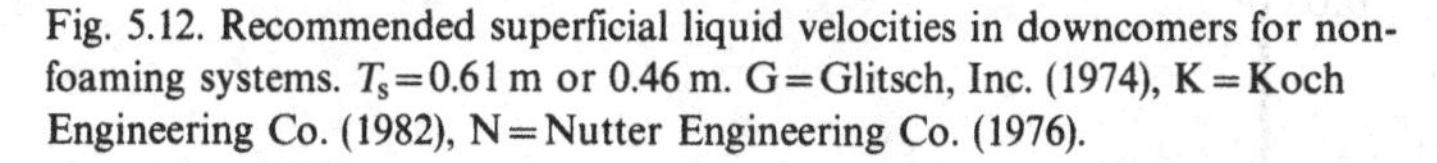

Fig. 5.12. Recommended superficial liquid velocities in downcomers for non-foaming systems. $T_s = 0.61$ m or 0.46 m. G = Glitsch, Inc. (1974), K = Koch Engineering Co. (1982), N = Nutter Engineering Co. (1976).

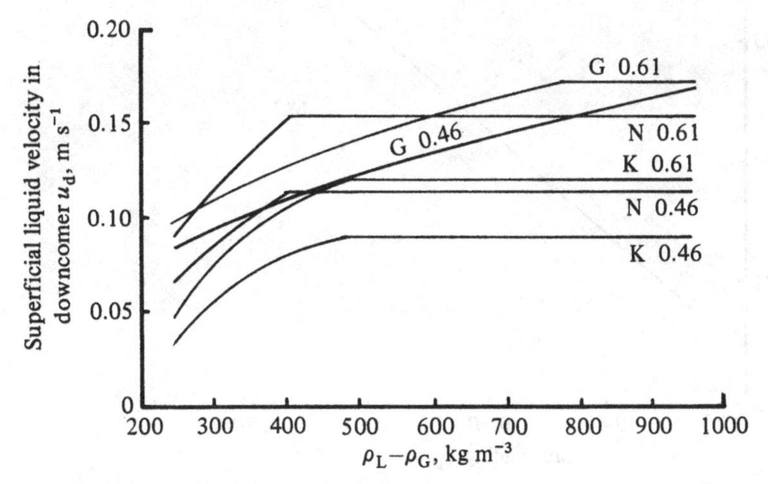

when the above effects become sufficiently serious to warrant the extra expense of more passes. Fig. 5.13 shows the maximum weir loads recommended by Nutter Engineering Co. (1976) as a function of tray spacing. MD trays take this to its logical conclusion by using a large number of downcomers, thereby keeping weir loads down to, typically, 0.006 $m^3 m^{-1} s^{-1}$. By so doing, tray spacings as low as 0.25 m can be used.

Picket, castellated or notched exit weirs are generally recommended for weir loads below about 0.0012 $m^3 m^{-1} s^{-1}$. Otherwise, at such low weir loads, the crest over the weir is very low and any out-of-levelness of the weir can cause non-uniform liquid flow across the tray and lead to a loss of tray efficiency. Inlet weirs are also frequently used at low liquid rates to ensure a positive downcomer seal.

5.4.3 *Choking of the downcomer mouth*

When a large liquid flow rate has to pass through the mouth of the downcomer, the latter can act as a restriction to flow and so require an increase in the froth height on the tray in order to achieve the required flow rate. The mouth of the downcomer becomes choked and normal weir flow is prevented. A critical liquid velocity, u_{dc}, exists (based on unaerated liquid) above which choking occurs (Lockett & Gharani 1979, Vikhman *et al.* 1976).

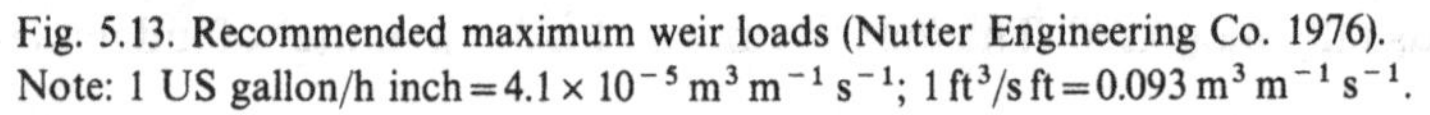

Fig. 5.13. Recommended maximum weir loads (Nutter Engineering Co. 1976). Note: 1 US gallon/h inch = 4.1×10^{-5} $m^3 m^{-1} s^{-1}$; 1 ft^3/s ft = 0.093 $m^3 m^{-1} s^{-1}$.

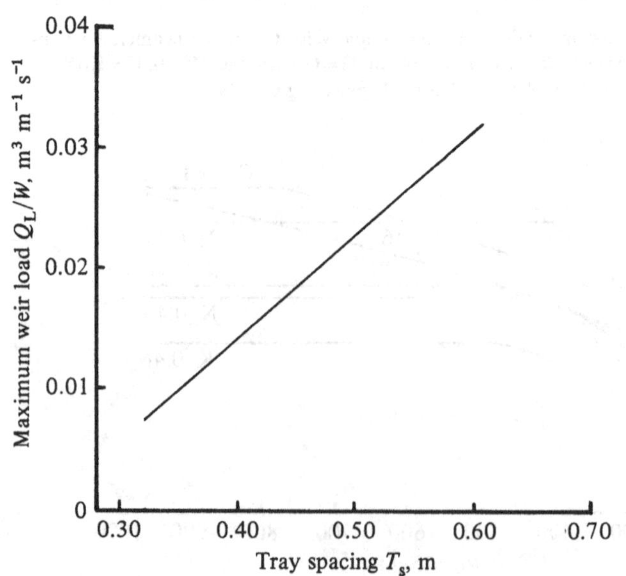

For air–water, u_{dc} varies between 0.35 and 0.48 m s^{-1} depending on the length of the exit calming zone. The key is the volume flow rate of froth into the mouth of the downcomer rather than the flow rate of unaerated liquid and the former depends on the extent of disengagement which occurs over the calming zone. Downcomer choking could explain reported observations that correlations for h_{cl} underpredict for heavily liquid loaded single-pass trays operating at high pressures (Bentham & Darton 1983).

5.4.4 *Clearance under the downcomer*

Some designers prefer always to set the clearance under the downcomer h_1 to be 12 mm less than the exit weir height. This ensures a positive downcomer seal under all circumstances. However, because of the crest over the weir and hydraulic gradient, a positive seal is usually achieved even when h_1 is somewhat higher than the weir height. A less-restrictive recommendation is to set h_1 such that the velocity of unaerated liquid under the downcomer is 0.5 m s^{-1}, which corresponds to $h_{udc} \approx 42$ mm. The preferred approach is to match the liquid velocity under the downcomer to that on the tray after aeration is established. This helps to ensure a smooth introduction of liquid onto the tray. When a large liquid turndown ratio is required, extra care is necessary to ensure that the downcomer will always remain sealed.

5.4.5 *Minimum weir length*

When liquid loads are low in comparison with vapour loads, such as in vacuum distillation, the downcomer area required to handle the liquid is often very small. In these cases it is usual not to allow W/D to fall below 0.6, even if the downcomer is then somewhat oversized. By so doing, the severest detrimental effects of liquid channelling across the tray can be avoided; see Section 9.13.

5.5 Oscillations at high vapour flow rates

The vapour–liquid dispersion on a tray can oscillate in a direction perpendicular to the direction of liquid flow in small-diameter columns operated at high vapour rates (Hinze 1965, Biddulph & Stephens 1974, Biddulph 1975*b*, Pinczewski & Fell 1975, Haug 1976, Scali & Zanelli 1982). Two types of oscillation have been identified as shown in Fig. 5.14. Biddulph & Stephens (1974) have proposed a dimensionless group B_s which can be used to predict the onset of oscillations, where

$$B_s = \frac{u_s De\, h_f \rho_G}{g D^3 \rho_L \alpha} \tag{5.23}$$

Fig. 5.14. Oscillations on trays. Views perpendicular to the direction of liquid flow.

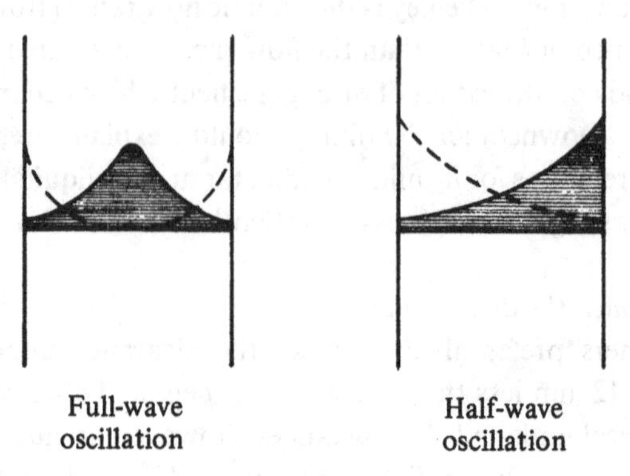

In eqn. (5.23), the eddy diffusivity is determined from the correlation of Barker & Self; see Section 9.7. Full-wave oscillation occurs for $B_s > 0.5 \times 10^{-5}$ and half-wave oscillation for $B_s > 2.5 \times 10^{-5}$. The onset of oscillations considerably increases both entrainment and weeping. Using eqn. (5.23) it is found, fortunately, that oscillations are unlikely to be significant in industrial columns having diameters in excess of about 1 to 1.5 m, depending on the pressure of operation. Nevertheless, the possible presence of oscillations in small laboratory columns does cast doubt on the relevance of hydraulics and efficiency studies in small columns, particularly in the spray regime, to larger industrial columns in which oscillations do not normally occur.

6

Weeping

Here we consider prediction of the onset and the rate of weeping. Its effect on tray efficiency is dealt with in Section 9.11.

6.1 Turndown ratio

If $(M_G)_{min}$ is the vapour flow rate below which acceptable tray efficiency is not obtained, then

$$\text{turndown ratio} = (M_G)_{min}/(M_G)_{at\ design}$$

Sometimes $(M_G)_{at\ flooding}$ is substituted for $(M_G)_{at\ design}$. Usually, in distillation, turndown ratio is defined for a constant ratio of vapour and liquid flow rates at design and under turndown conditions.

For sieve trays, in vacuum distillation, the turndown ratio is typically 0.7, but at higher pressures it can be as low as 0.2. Since the 'acceptable' tray efficiency is vague and varies from case to case, so turndown ratio is inherently imprecise.

Weeping occurs when the pressure drop of the vapour passing through the tray deck is insufficient to support the liquid. Consequently, contrary to popular belief, sieve trays can be designed for very low turndown ratios if sufficient dry tray pressure drop is provided at the lowest vapour flow rates used. This is accomplished by using a small fractional perforated area. The difficulty is that an excessively high pressure drop is then obtained at design conditions. In vacuum distillation this is usually unacceptable as it raises the pressure and hence temperature in the reboiler, which can lead to product degradation. Even at medium to high pressures problems can occur, such as the requirement for increased tray spacing to accommodate the higher downcomer backup at design flow rates. The possibility of spray regime operation and excessive entrainment at design conditions is another point to consider. The actual turndown ratio provided is usually a compromise between these various factors.

Valve trays maintain a higher dry tray pressure drop than sieve trays at low vapour flow rates because of the kink in the pressure-drop curve caused by the valves (Fig. 4.5). For the same pressure drop at design, a valve tray can therefore usually achieve a somewhat lower turndown ratio than a sieve tray.

Apart from the strategy described above of increasing the dry tray pressure drop to increase the turndown ratio, an alternative approach is to reduce the amount of liquid on the tray under turndown conditions and so reduce the tendency to weep. Zanelli (1975) described a modified weir design for sieve trays where a portion of the liquid was allowed to flow through a horizontal slot part way up the weir. This gave the weir some of the characteristics of an orifice for liquid flow, and tended to reduce the clear liquid height on turndown to at least partly follow the reduction in the dry tray pressure drop. Zanelli claimed that weeping was reduced using the device.

6.2 Weep point

6.2.1 *Definition of weep point*

Fig. 6.1 shows how the weep rate typically varies with hole gas velocity. A higher exit weir increases the clear liquid height and also increases the weep rate. The curves do not intersect the axis sharply and the hole gas velocity where weeping starts, the weep point, is difficult to locate by measuring the weep rate. A graphical weep point (GWP) can be defined

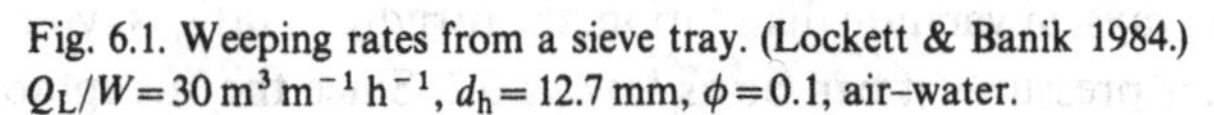

Fig. 6.1. Weeping rates from a sieve tray. (Lockett & Banik 1984.)
$Q_L/W = 30\ \mathrm{m^3\,m^{-1}\,h^{-1}}$, $d_h = 12.7$ mm, $\phi = 0.1$, air–water.

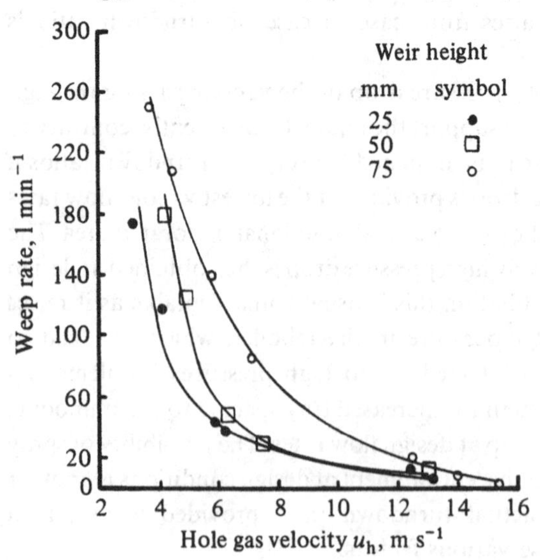

from the change in slope of the pressure drop curve at constant liquid rate; see Fig. 6.2. A log–log plot is often used to aid in locating the GWP. The measured weep point and the GWP often do not agree, particularly for trays of high fractional open area (> 15%).

It is generally accepted (Prince 1960 and Prince & Chan 1965) that below the weep point some holes pass liquid while the remainder continue bubbling. In fact, visual observation shows that a given hole randomly switches between either weeping or bubbling. The average number of holes devoted to each activity, however, remains constant at fixed gas and liquid flow rates. Prince & Chan (1965) have developed an equation to predict the fraction of holes which bubble.

Included on Fig. 6.2 is the dry tray pressure drop (valid strictly only at and above the weep point when all holes bubble). The ratio

$$Fr_h = u_h \left[\frac{\rho_G}{g h_{cl}(\rho_L - \rho_G)} \right]^{0.5} \tag{6.1}$$

is also shown, where h_{cl} has been estimated from $h_{cl} = h_{WT} - h_{DT}$. Zuiderweg (1982) has pointed out that ξ in eqn. (4.8) is usually close to 2, so that

Fig. 6.2. Sieve tray pressure drop during weeping. (Banik 1982, Dhulesia 1980.) $d_h = 12.7$ mm, $h_w = 50$ mm, $\phi = 0.1$, air–water. Parameter is liquid weir load Q_L/W m^3 m^{-1} h^{-1}.

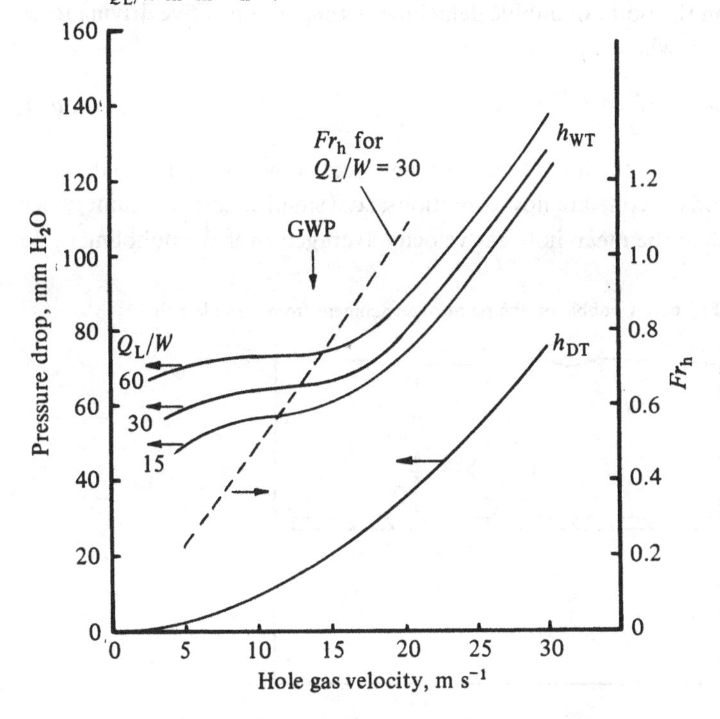

$Fr_h \approx (h_{DT}/h_{cl})^{0.5}$. For the particular data shown, $Fr_h \approx 0.7$ at the weep point. This is taken up again in Section 6.2.3.

6.2.2 *Theoretical prediction of the weep point*

McCann & Prince (1969) have predicted the onset and rate of weeping during bubble formation from a single hole based on the theory outlined in Section 2.2.2. A simplified treatment due to Kupferberg & Jameson (1970) is outlined below.

After a bubble has left the hole, and before the next one has started to form, weeping is possible if the liquid pressure at the tray surface is higher than the pressure in the chamber below the hole. Fig. 6.3 shows in a simplified way how this can occur by considering a spherical bubble at the point of detachment.

The pressure in the bubble is equal to the hydrostatic pressure at its centre. The chamber pressure is the sum of the pressure in the bubble and the pressure drop across the hole at detachment. Relative to the pressure at the liquid surface, in terms of liquid head

$$\text{chamber pressure} = y + \frac{\xi \rho_G u_h^2}{2\rho_L g} \tag{6.2}$$

$$\text{pressure in liquid at tray surface} = y + r \tag{6.3}$$

So, just on the point of bubble detachment, there is a positive driving force for weeping when

$$r > \frac{\xi \rho_G u_h^2}{2\rho_L g} \tag{6.4}$$

The maximum likelihood of weeping occurs when u_h is small. This corresponds to constant flow conditions, i.e. a small chamber volume, when u_h is close to the mean hole gas velocity averaged over the bubbling cycle.

Fig. 6.3. A bubble on the point of detachment from a single hole.

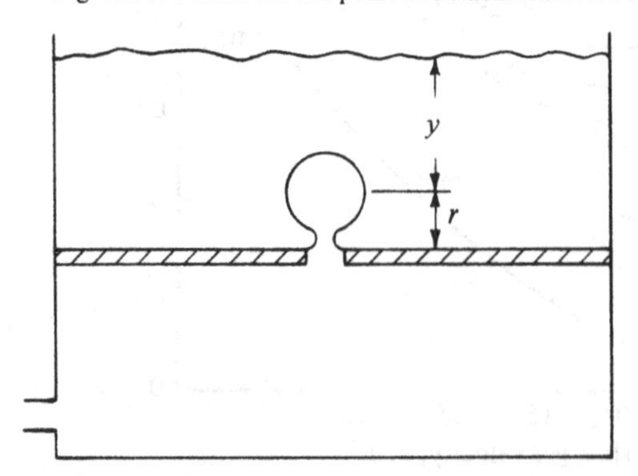

For a large chamber volume, the chamber pressure remains relatively constant because of damping so that just on the point of bubble detachment u_h is rather higher than its mean value. Consequently, the tendency to weep is greatest for small chamber volumes. This was confirmed experimentally by McCann & Prince (1969).

Kupferberg & Jameson used eqn. (2.3) to estimate r at detachment. Substituting into eqn. (6.4) gives for weeping

$$\xi \frac{\rho_G}{\rho_L}\left(\frac{u_h^2}{gd_h}\right)^{0.8} < 1.25 \tag{6.5}$$

It was found that the weep-point data of McCann & Prince could be correlated quite well using a constant of 1.27.

Neither the simple theory given above nor more-detailed theories (McCann & Prince 1969) predicts any influence of the clear liquid height on the weep point. Experiment shows, however, that there is indeed an influence. More importantly, theories based on bubbling from single holes have so far not been extended to multihole bubbling except to a very limited extent. There are two major difficulties. The pressure field in the liquid around a given bubble is altered by the presence of neighbouring bubbles. Also it is difficult to determine the effective chamber volume per hole in a multihole situation. The usual approach for the latter has been to assume the chamber volume associated with each hole is the total chamber volume divided by the total number of holes (Kupferberg & Jameson 1970, Tadaki & Maeda 1963). This implies that multihole plates are associated with very small chamber volumes per hole and operate under approximately constant flow conditions. The difficulty is that the individual volumes associated with each hole are not isolated from each other. Any pressure fluctuations beneath a given hole will tend to be damped by flow of gas from adjacent areas. It seems the opposite assumption is more likely – the pressure below a multihole plate is approximately constant and independent of the random formation of bubbles from individual holes. In other words, multihole plates are more likely to operate in the constant pressure regime and to weep less than single holes. Haselden & Thorogood (1964) and more recently Miyahara *et al.* (1983*b*) have published some results on bubble sizes on formation which tend to confirm this view. If the appropriate regime is indeed constant pressure, then the driving force for weeping is much reduced. In this case, surges of excess pressure in the liquid, such as those found in the wake of a rising bubble as discussed by Jameson & Kupferberg (1967), then appear to be the most probable source of the driving force for weeping.

In any event, the simple theory outlined above breaks down. It was hoped at the time these detailed studies of bubbling from single holes were carried out that they would ultimately be extended to multiple holes. Unfortunately, so far, this has not happened.

A slightly different theoretical approach was adopted by Ruff *et al.* (1978), who considered the stability of the gas jet connecting a growing bubble to the hole. To ensure no weeping it was suggested that, for small holes,

$$\frac{\rho_G d_h u_h^2}{\sigma} > 2.0 \quad \text{(2.7 for safety)} \tag{6.6}$$

and for large holes,

$$\frac{u_h^2}{g d_h}\left(\frac{\rho_G}{\rho_L - \rho_G}\right)^{1.25} > 0.37 \quad \text{(0.75 for safety)} \tag{6.7}$$

The transition between 'small' and 'large' holes is given by

$$d_h = 2.32\left(\frac{\sigma}{\rho_G g}\right)^{0.5}\left(\frac{\rho_G}{\rho_L - \rho_G}\right)^{0.625} \tag{6.8}$$

Except in cryogenic applications the holes used in sieve trays are usually 'large'. Note that eqn. (6.7) with a constant of 0.56 is identical to eqn. (6.5) with $\xi = 2$.

A third semi-theoretical approach can be derived from work by Wallis & Kuo (1976). They studied the critical conditions for penetration of liquid from a reservoir into the top open end of a vertical tube, with air flowing up the tube. This can loosely be identified with the critical conditions for the onset of weeping on a sieve tray. Their results for tubes having diameters corresponding to typical sieve tray hole diameters can reasonably be represented by

$$\frac{u_h^2}{g d_h}\left(\frac{\rho_G}{\rho_L - \rho_G}\right) \approx 0.25 \tag{6.9}$$

Eqn. (6.9) also bears a strong resemblance to eqns. (6.5) and (6.7).

6.2.3 *Empirical correlations for the weep point*

All empirical weep-point correlations involve the F factor based on the hole gas velocity. They also generally include h_{cl} and most attach less importance to d_h than do the theoretical predictions above.

One of the earliest correlations is that of Fair (1963), which was based on data obtained by Mayfield *et al.* (1952) and Hutchinson *et al.* (1949). A good approximation to Fair's correlation is

$$h_{DT} + h_R = 0.081(h_{cl})^{0.61} \quad 0.06 < \phi < 0.14 \tag{6.10}$$

Neglecting h_R and using eqn. (4.8), this can be rearranged to give

$$Fr_h = u_h\left(\frac{\rho_G}{g\rho_L h_{cl}}\right)^{0.5} = \frac{0.40}{\xi^{0.5}h_{cl}^{0.2}} \tag{6.11}$$

Note that, in Fair's correlation, h_{cl} is determined from the sum of the weir height and the crest of unaerated liquid flowing over the weir calculated from Francis's formula.

However, in the paper of Mayfield *et al.* (1952) sufficient information is given to estimate h_{cl} as the difference between the total and dry tray pressure drops. When this is done, Mayfield's own correlation of his data is well represented by

$$Fr_h = u_h\left(\frac{\rho_G}{gh_{cl}\rho_L}\right)^{0.5} = 0.45\left(\frac{2}{\xi}\right)^{0.5}, \quad d_h = 0.0048 \text{ m}$$

For the tray used, ξ varied between 1.78 and 1.13, giving

$$Fr_h = 0.48 \text{ to } 0.60$$

Eduljee's (1972) weep-point correlation of most of the early data is

$$u_h\rho_G^{0.5} = 6.71 + (336\,d_h + 0.305)(39.4\,h_{cl} + 1.1),$$
$$0.002 \text{ m} < d_h < 0.008 \text{ m} \tag{6.12}$$

The strong dependence on hole diameter was arrived at by selectively omitting those data points which did not show that $u_h\rho_G^{0.5}$ increased with d_h. As such it must be considered of doubtful validity.

Zuiderweg (1982) used the data of Mayfield *et al.* and of FRI (Sakata & Yanagi 1979) and proposed the following correlations for the weep point:

mixed-froth/free bubbling regime

$$Fr_h = 1 - \frac{0.15\,Q_L}{Wu_s h_{cl}}\left(\frac{\rho_L}{\rho_G}\right)^{0.5} \tag{6.13}$$

emulsion regime

$$Fr_h = 0.45 \tag{6.14}$$

One difficulty with all these correlations for the weep point is that h_{cl} was usually not reported and the proposed correlation depends strongly on how h_{cl} was estimated. Zuiderweg used eqn. (3.26) to estimate h_{cl} for the FRI data. At very low liquid flow rates h_{cl} tends to zero independently of the weir height using eqn. (3.26), and Fr_h is therefore overestimated. When h_{cl} is estimated from $(h_{WT} - h_{DT})$ for the visually observed weep-point data of FRI, the values of Fr_h obtained are shown in Fig. 6.4. Also shown in Fig. 6.4 are values of Fr_h obtained by Zanelli (1975), where h_{cl} at the weep point was determined directly by manometers set into the tray floor. Note that Fig. 6.4 shows no evidence of a break point at the transition to emulsion flow as reported by Zuiderweg. The data shown can be correlated by $Fr_h = 0.68 \pm 0.12$, and this is only slightly different from the value of $Fr_h = 0.48-0.60$ obtained from Mayfield's data. Although superficially this appears satisfactory it should be realised that there is still considerable scatter in the data. Fig. 6.2 shows how Fr_h typically changes with a reduction of column loading, and it is evident that the uncertainty in predicting the weep point is considerable.

6.3 Empirical correlations for the weep rate

In comparison with the weep point, there have been fewer studies where the weep rate has been measured (Brown 1958, Wada *et al.* 1966, Zenz *et al.* 1967, Lemieux & Scotti 1969, Nutter 1972, 1979, Koziol & Koch 1973, 1976, Brambilla *et al.* 1979, Zhou *et al.* 1980, Lockett & Banik 1984). Weep-rate correlations have been given by Wada for 3 mm holes, and by Zenz for 1.6–6.4 mm holes. Zenz correlated the weep rate per unit hole area (weep flux) as a function of the difference between the actual hole F_h factor ($u_h \rho_G^{0.5}$) and the hole F_h factor at the weep point. This places a premium on having an accurate correlation for the latter but, as discussed above, a sufficiently accurate correlation is not available.

In an attempt to avoid this problem, Lockett & Banik (1984) correlated

Fig. 6.4. Correlation of weep point data.

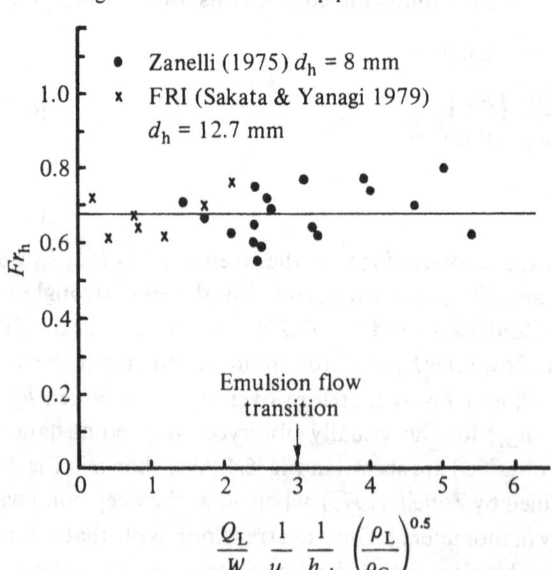

the weep flux directly as a function of Fr_h. Their correlation is shown in eqn. (6.15):

$$\text{weep flux} = \frac{\text{weep rate (m}^3\,\text{s}^{-1})}{\text{total hole area (m}^2)} = 0.020\,Fr_h^{-1} - 0.030 \qquad (6.15)$$

The hole gas velocity used for Fr_h was based on all holes bubbling and the clear liquid height was estimated from $(h_{WT} - h_{DT})$. For design purposes it was recommended that h_{cl} should be determined from Colwell's correlation – eqn. (3.17). As the latter involves the liquid load over the weir, which in turn depends on the weep rate, an iterative calculation is required. Eqn. (6.15) is not inconsistent with the weep point correlations discussed above since at zero weep flux

$$Fr_h = 0.67 \qquad (6.16)$$

It was found that a tray having $\phi = 0.2$ wept slightly more than the correlation given in Fig. 6.5, and a tray having $d_h = 3$ mm wept slightly less. The former was attributed to severely non-uniform conditions on the tray and the latter to a film of liquid which collected on the underside of a tray having small-diameter holes. Both phenomena are difficult to deal with quantitatively. It was found that a surface tension variation over the range 23–72 mN m^{-1} had an undetectable influence on the rate of weeping.

Fig. 6.5. Weep flux correlation for sieve trays. (Lockett & Banik 1984.) Air–water, air–hydrocarbon oil (Isopar M). $d_h = 3$–12.7 mm, $\phi = 0.1$–0.2, $\rho_G = 1.2$ kg m^{-3}, $\rho_L = 780$–1000 kg m^{-3}, $\sigma = 23$–72 mN m^{-1}.

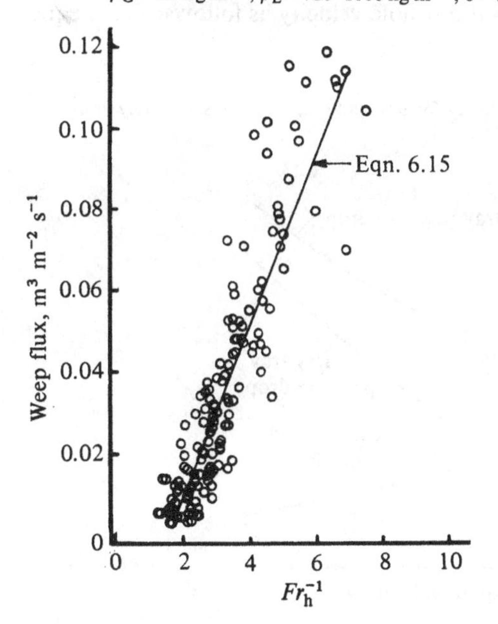

6.3.1 *Dump point*

The dump point is the gas velocity at which the weep fraction is just 1.0, so that no liquid flows over the weir. Correlations have been given by Prince & Chan (1965) and by Brambilla *et al.* (1979). The dump point can also be estimated using a weep flux correlation such as that given in Fig. 6.5. In practice the dump point is generally of little concern, since under these conditions tray efficiency is low and difficult to estimate. Normally, if a column needed to be operated at such a low vapour flow rate, consideration would be given to retraying or at least blanking off part of the tray deck.

6.4 Oscillation at low vapour flow rates

There have been a number of reports of flow-induced vibrations occurring under turndown conditions which in some cases have led to mechanical damage to the trays (Brierley *et al.* 1979, Priestman & Brown 1981). Priestman has suggested that there is a critical F_h factor for which vibrations can occur. For air–water the critical F_h factor is $14.1 \text{ kg}^{0.5} \text{ m}^{-0.5} \text{ s}^{-1}$ and for hydrocarbons it is slightly lower. Brierley has suggested that the onset of vibrations may be associated with a condition shown in Fig. 6.6 where the wet tray pressure drop passes through a minimum as the hole vapour velocity varies. It is postulated that near the minimum the hole vapour flow rate can oscillate between two values, both of which are associated with the same wet tray pressure drop. We can derive an expression for the critical vapour hole velocity as follows. From eqns.

Fig. 6.6. Critical vapour velocity for flow-induced vibrations. (Brierley *et al.* 1979.)

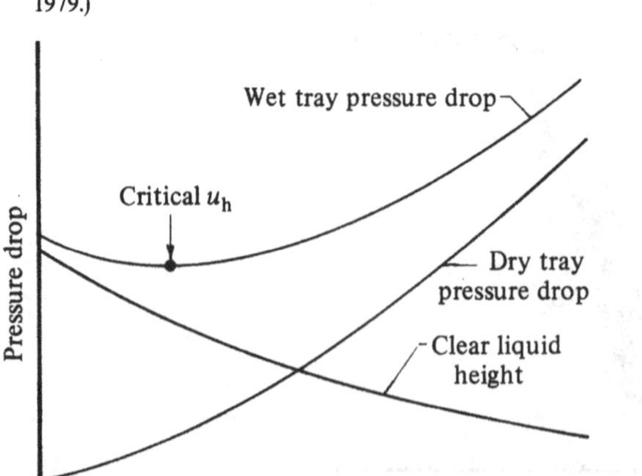

(3.26) and (4.8), and neglecting residual pressure drop,

$$h_{WT} = \frac{\xi \rho_G u_h^2}{2g\rho_L} + 0.6\left(\frac{Q_L p h_w^2}{W\phi u_h}\right)^{0.25}\left(\frac{\rho_L}{\rho_G}\right)^{0.125} \tag{6.17}$$

Using the condition that $\partial h_{WT}/\partial u_h = 0$ at the critical velocity, it follows that:

$$(F_h)_{crit} = (u_h \rho_G^{0.5})_{crit} = \left(\frac{0.6\,g}{\xi}\right)^{1/(2.25)}\left(\frac{Q_L p h_w^2}{W\phi}\right)^{1/9}(\rho_L)^{0.5} \tag{6.18}$$

For the following typical values of the parameters involved in eqn. (6.18): $\xi = 2, Q_L/W = 0.01\ \mathrm{m^3\,m^{-1}\,s^{-1}}, \phi = 0.1, p = 38 \times 10^{-3}\ \mathrm{m}, h_w = 50 \times 10^{-3}\ \mathrm{m}$, we have

$$(F_h)_{crit} = 0.447(\rho_L)^{0.5} \tag{6.19}$$

For water, $(F_h)_{crit} = 14.1\ \mathrm{kg^{0.5}\,m^{-0.5}\,s^{-1}}$, from eqn. (6.19), which is in remarkable agreement with Priestman's experimental result. In addition, because of the small exponent on the term $(Q_L p h_w^2/W\phi)$ in eqn. (6.18), $(F_h)_{crit}$ is relatively independent of tray design parameters. In agreement with Priestman, $(F_h)_{crit} = 14.1$ represents an upper limit and hydrocarbons should have a lower value of $(F_h)_{crit}$ because of their lower value of ρ_L.

Critical conditions can be avoided by reducing ϕ so as to ensure that $F_h > (F_h)_{crit}$ even at turndown conditions. The penalty is an increased tray pressure drop and increased downcomer backup at design conditions. An alternative is to provide extra stiffening of the trays and to avoid integrally formed support beams as suggested by Brierley.

7

Tray efficiency

7.1 The use of efficiencies in column design

A first step in deciding how many actual trays to put into a column to achieve a required separation is to use a column simulation computer program. These usually deal with theoretical trays. They require as input the total number of theoretical trays in the column and the location of feeds and side draws, again in terms of theoretical trays.

Most programs allow the top and bottom product purities and recoveries to be fixed for the key components and they determine the required reflux ratio corresponding to the assumed number of theoretical trays. In this way a 'number of theoretical trays versus reflux ratio' curve can be generated which can be used to decide on the appropriate combination of these two variables to be used in the design. Apart from the obvious utility costs versus investment costs tradeoffs, other factors often play a part in the decision. Some examples are:

- availability of 'free' utilities by appropriate energy integration with the rest of the plant;
- maximum column height limitations;
- sensitivity of column performance to uncertainty in vapour–liquid equilibrium relationships and in tray efficiency. In the latter case it is difficult to compensate for a shortfall in tray efficiency during column operation by increasing the reflux ratio if the design point is close to total reflux. The reverse is true near minimum reflux.

A discussion of column simulation (often called tray counting) is outside the scope of this book. Details may be found in texts by Holland (1963, 1966, 1975), Henley & Seader (1981) and King (1980). In practice, most companies regularly involved with distillation have in-house column simulation computer programs or have access to one of the numerous programs offered commercially.

After column simulation, a section efficiency E_0 is usually applied over a group of theoretical trays N_T to determine the number of actual trays N_A in that section of the column, where

$$N_A = N_T/E_0 \tag{7.1}$$

The size of a section depends on the judgement of the designer. In the crudest designs, the whole column is taken as one section and E_0 is then called the overall column efficiency. For a single feed column, with products withdrawn only at the top and bottom, two sections are often adequate corresponding to the trays above and below the feed. More sections should be used where multiple feeds or side draw streams are involved. Quite often, trays of different design are used in different parts of the column to accommodate varying hydraulic loads and this should also influence the choice of column sections for efficiency estimation. The usual procedure is to locate a tray which is representative of conditions in a given section and to estimate a section efficiency based on conditions on that tray.

Section 8.1 gives an outline procedure for calculation of section efficiencies for binary mixtures, and Section 10.2 addresses the problems of using section efficiencies with multicomponent mixtures.

Some column simulation programs deal with actual trays. The user has to specify a tray efficiency for each component on each tray. Such programs are to be preferred in general, and particularly when the location of side draw streams is being decided.

Various definitions of efficiency have been proposed over the years, but it seems the more rigorous and acceptable the definition, the more difficult it is to use practically. For example, the section efficiency, E_0, is easy to use but is meaningless for multicomponent mixtures, whereas the Standart efficiencies (Section 7.5) are the soundest fundamentally but apparently have never been used in an actual design. Some proposed definitions of tray efficiency are reviewed in what follows together with their attributes and disadvantages.

7.2 Murphree tray efficiencies

The Murphree tray efficiency (Murphree 1925) is defined by comparing a real tray shown in Fig. 7.1*a* with an ideal tray. Fig. 7.1*b* shows the ideal tray used for the Murphree vapour phase tray efficiency, E_{MV}, and Fig. 7.1*c* shows the ideal tray used for the Murphree liquid phase tray efficiency, E_{ML}. Note that in each case one of the entering streams differs in composition between the real tray and the ideal tray (denoted by superscript $''$). For each component

$$E_{MV} = \frac{\bar{y}_n - \bar{y}_{n-1}}{y_n^* - \bar{y}_{n-1}} \tag{7.2}$$

$$E_{ML} = \frac{\bar{x}_n - \bar{x}_{n+1}}{x_n^* - \bar{x}_{n+1}} \tag{7.3}$$

where

$$y_n^* = m\bar{x}_n + b \tag{7.4}$$

and

$$\bar{y}_n = mx_n^* + b \tag{7.5}$$

Eqn. (7.4) defines a vapour composition y_n^* which is in equilibrium with the exit liquid composition $\bar{x}_n$, where the exit liquid is assumed to be at its bubble point in both Figs. 7.1*a* and *b*. Similarly, eqn. (7.5) defines a liquid composition x_n^* which is in equilibrium with the exit vapour composition $\bar{y}_n$, where the exit vapour in Figs. 7.1*a* and *c* is assumed to be at its dew point. (Note: in general, *m* and *b* are functions of temperature, pressure and also composition.)

7.2.1 *Relationship between* E_{ML} *and* E_{MV}

If the operating line is straight (constant *G* and *L*), a component mass balance over the real tray gives

Fig. 7.1. Nomenclature for definitions of tray efficiency. (*a*) Real tray. (*b*) Ideal tray for definition of E_{MV}. (*c*) Ideal tray for definition of E_{ML}. (*d*) Ideal tray for definition of E_H.

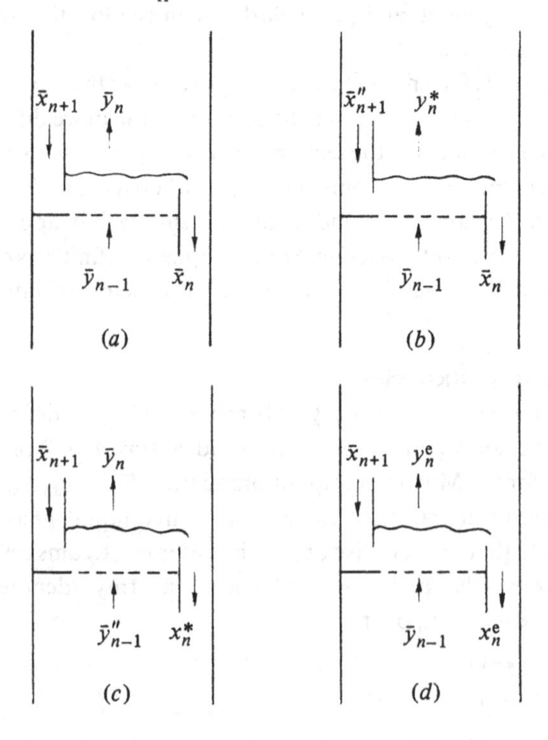

$$L(\bar{x}_{n+1}-\bar{x}_n)=G(\bar{y}_n-\bar{y}_{n-1}) \tag{7.6}$$

If the equilibrium line is also straight (m and b independent of composition over the composition range involved), eqns. (7.4) and (7.5) give

$$m=\frac{y_n^*-\bar{y}_n}{\bar{x}_n-x_n^*} \tag{7.7}$$

so that

$$\lambda=\frac{mG}{L}=\left(\frac{y_n^*-\bar{y}_n}{\bar{x}_n-x_n^*}\right)\left(\frac{\bar{x}_{n+1}-\bar{x}_n}{\bar{y}_n-\bar{y}_{n-1}}\right) \tag{7.8}$$

From eqns. (7.2), (7.3) and (7.8)

$$\lambda\left(\frac{1}{E_{ML}}-1\right)=\left(\frac{1}{E_{MV}}-1\right) \tag{7.9}$$

or

$$E_{MV}=\frac{E_{ML}}{E_{ML}+\lambda(1-E_{ML})} \tag{7.10}$$

Murphree tray efficiencies are very convenient for tray-to-tray calculations involving a binary mixture. E_{MV} is used when moving up the column and E_{ML} when moving down.

7.2.2 *Relationship between* E_{MV} *and* E_0

Lewis (1936) derived the relationship between E_{MV} and the section efficiency E_0 as

$$E_0=\frac{\ln[1+E_{MV}(\lambda-1)]}{\ln\lambda} \tag{7.11}$$

Eqn. (7.11) is based on the assumption of straight operating and equilibrium lines and a constant value of E_{MV} from tray to tray. These assumptions are often appropriate only for difficult separations of low relative volatility when $\lambda\approx 1.0$ and $E_0\approx E_{MV}$.

7.3 Hausen tray efficiency (Hausen 1953)

One objection to the Murphree efficiency is that the entering streams differ for the real and ideal trays. A more satisfying definition of tray efficiency is obtained when the entering streams are held constant and both exit streams change according to the efficiency of contact on the tray. The Hausen tray efficiency is defined using this approach. Fig. 7.1*d* shows the ideal tray used in the definition of the Hausen efficiency, which is compared with the real tray of Fig. 7.1*a*. In Fig. 7.1*d* the exit streams are in equilibrium so that for each component

$$y_n^e=mx_n^e+b \tag{7.12}$$

Hausen tray efficiencies are defined for each component as

$$E_{HV}=\frac{\bar{y}_n-\bar{y}_{n-1}}{y_n^e-\bar{y}_{n-1}} \quad \text{and} \quad E_{HL}=\frac{\bar{x}_n-\bar{x}_{n+1}}{x_n^e-\bar{x}_{n+1}} \tag{7.13}$$

A component mass balance on the tray shown in Fig. 7.1*d* gives

$$L(\bar{x}_{n+1}-x_n^e)=G(y_n^e-\bar{y}_{n-1}) \tag{7.14}$$

Dividing eqn. (7.6) by eqn. (7.14) shows that

$$E_H=E_{HV}=E_{HL} \tag{7.15}$$

So the Hausen efficiency has the additional advantage over the Murphree efficiency that a single efficiency suffices for both phases.

7.3.1 Relationship between E_H and E_{MV}

The compositions involved in Figs. 7.1*a–d* are shown in Fig. 7.2 for a binary mixture:

$$E_H=\frac{BF}{CH}, \quad E_{MV}=\frac{BF}{DF}, \quad E_{ML}=\frac{AF}{AG}$$

Now

$$E_H=\frac{BF}{DF}\cdot\frac{DF}{CH}=E_{MV}\left[\frac{BF}{CH}+\frac{DB}{CH}\right]$$

Fig. 7.2. Representation of E_H, E_{MV} and E_{ML} for a binary mixture.

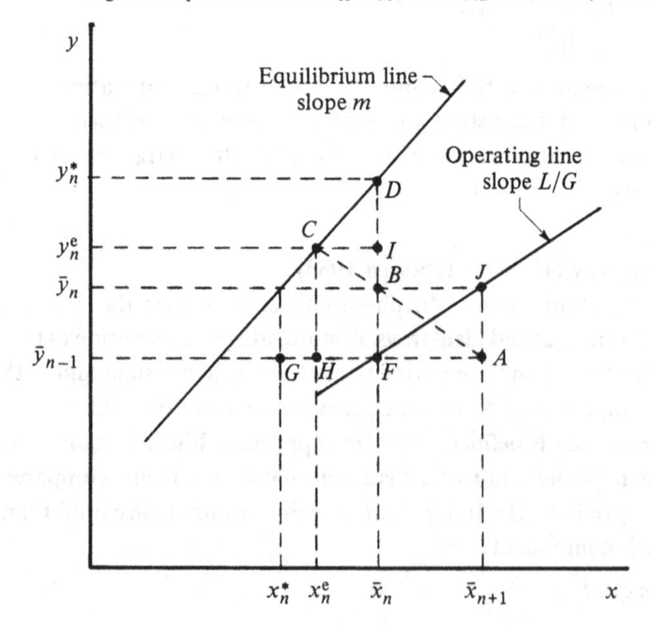

But

$$DB = DI + IB$$
$$= m \cdot CI + IB$$
$$= m \cdot \frac{G}{L} \cdot IB + IB$$
$$= (1+\lambda)IB$$
$$= (1+\lambda)(CH - BF)$$

so that

$$E_H = E_{MV}[E_H + (1+\lambda)(1-E_H)]$$

and

$$E_H = \frac{E_{MV}(1+\lambda)}{1+\lambda E_{MV}} \tag{7.16}$$

7.4 Phase temperatures and saturation

Standart (1965) pointed out that the convention adopted in the Murphree definitions, that the exit streams are saturated, is a purely arbitrary assumption. Since these streams are not in composition equilibrium, the assumption of saturation implies that they are at different temperatures. Another, equally plausible, assumption is that they are at the same temperature but unsaturated. A third, and more likely, situation is that the exit streams are neither saturated nor in thermal equilibrium. In fact, the Murphree and Hausen efficiencies provide no information about the actual thermal state of the exit streams.

When dealing with binary or pseudo-binary mixtures with constant G and L this generally causes no difficulties. It is of more concern when tray-to-tray calculations for real trays involve enthalpy balances or when dealing with multicomponent mixtures. Sargent and Murtagh (1969) have shown how Murphree efficiencies may be adapted for the latter case. Since they worked in terms of vapour compositions, their procedure was to define y_n^* in eqn. (7.2) as the vapour composition in equilibrium with the liquid leaving the real tray if the latter were brought to its bubble-point temperature. The actual temperatures of the streams leaving a real tray were determined from thermal efficiencies similar to those proposed by Nord (1946):

$$E_{TV} = \frac{T_{Vn} - T_{Vn-1}}{T_{Vn}^* - T_{Vn-1}} \quad \text{and} \quad E_{TL} = \frac{T_{Ln} - T_{Ln+1}}{T_{Ln}^* - T_{Ln+1}} \tag{7.17}$$

T_{Vn}^* and T_{Ln}^* are the dew- and bubble-point temperatures, respectively, of the vapour and liquid leaving the real tray. Prediction of E_{TV} and E_{TL} was not considered by Sargent & Murtagh although it is likely they are fairly close to unity. If they are unity, of course, the exit streams are saturated and at different temperatures.

7.5 Standart efficiency

Apart from the question of thermal equilibration, Standart (1965) pointed out that both the Murphree and Hausen efficiencies assume constant G and L over the tray. He defined very general efficiencies which are not subject to the aforementioned qualifications. They are based on the real tray of Fig. 7.1*a* and the ideal tray of Fig. 7.1*d*:

overall material efficiency

$$E_S = \frac{G_n - G_{n-1}}{G_n^e - G_{n-1}} = \frac{L_n - L_{n+1}}{L_n^e - L_{n+1}} \tag{7.18}$$

component efficiencies

$$E_{Si} = \frac{G_n \bar{y}_n - G_{n-1}\bar{y}_{n-1}}{G_n^e y_n^e - G_{n-1}\bar{y}_{n-1}} = \frac{L_n \bar{x}_n - L_{n+1}\bar{x}_{n+1}}{L_n^e x_n^e - L_{n+1}\bar{x}_{n+1}} \tag{7.19}$$

enthalpy efficiency

$$E_{SH} = \frac{G_n H_n - G_{n-1} H_{n-1} + r_n Q_n}{G_n^e H_n^e - G_{n-1} H_{n-1} + r_n Q_n} = \frac{L_n h_n - L_{n+1} h_{n+1} + (1-r_n) Q_n}{L_n^e h_n^e - L_{n+1} h_{n+1} + (1-r_n) Q_n} \tag{7.20}$$

where superscript e refers to the exit streams from the ideal tray. Q_n is the rate of heat loss from the tray to the surroundings and r_n the fraction lost by the vapour. For equimolar overflow, eqn. (7.19) reduces to the Hausen efficiency.

The difficulty of using the Standart efficiencies (and to a lesser extent the Hausen efficiency) is that a flash calculation must be performed on each tray to determine the exit streams from the ideal tray. In this regard Murphree efficiencies are more convenient in spite of their lack of rigour.

7.6 Holland's vaporisation efficiency

Holland (1963) and Holland & McMahon (1970) have proposed a vaporisation tray efficiency, E_{ni}^V, which is of a quite different type than the tray efficiencies so far considered. The definition is

$$E_{ni}^V = \bar{f}_{ni}^V / \bar{f}_{ni}^L \tag{7.21}$$

where $\bar{f}^V_{ni}$ = fugacity of component i in the vapour stream leaving tray n evaluated at the conditions of this vapour stream, and $\bar{f}_{ni}^L$ = fugacity of component i in the liquid stream leaving tray n evaluated at the conditions of this liquid stream.

For the particular case that the exit vapour and liquid are at the same temperature and the vapour phase is an ideal solution we have

$$E^{V}_{ni} = \bar{y}_{ni}/(\gamma^{L}_{ni}K_{ni}\bar{x}_{ni}) \tag{7.22}$$

Here $(\gamma^{L}_{ni}K_{ni}\bar{x}_{ni})$ is evaluated at the conditions of the exit liquid stream, where, for component i, γ^{L}_{ni} is the activity coefficient, K_{ni} is the ideal solution K value and $\bar{x}_{ni}$ is the mole fraction.

The main attraction of the vaporisation efficiency is its usefulness in computation. It can be related to the Murphree efficiency (Holland & McMahon 1970, Holland 1980) so it can be calculated if the latter can be calculated. Its disadvantage is that its value varies over the column with changes in composition. Examination of eqn. (7.22) shows that at the top of the column, as $\bar{y}_{ni}$ and $\bar{x}_{ni}$ tend to 1.0, E^{V}_{ni} also tends to 1.0. Consequently, the numerical value of E^{V}_{ni} bears no direct relationship to the ease or difficulty of the separation. This places it at a fundamental disadvantage as a realistic measure of efficiency, and as such it has been severely criticised (Standart 1971, Medina *et al.* 1978).

7.7 Measurement of efficiency

In pilot-scale columns, measurement of the efficiency of individual components on a particular tray may be made by determining the compositions of the inlet and exit streams. As Murphree efficiencies involve the ratio of composition differences, considerable accuracy is required. Problems arise in avoiding partial vaporisation of liquid samples and in ensuring that vapour samples are dry (Kastanek & Standart 1967, Lockett & Ahmed 1983). If the relative volatility is constant, the overall column efficiency of a binary mixture can be determined from the terminal compositions at total reflux using the Fenske–Underwood equation. Silvey & Keller (1966, 1969) have shown how tray-by-tray composition samples can be used to improve the accuracy of this approach.

Determination of accurate efficiencies from industrial columns can be extremely difficult (Sealey 1970). Biddulph & Ashton (1977) have described a case history involving a benzene-toluene-xylene column, and they pointed out that plant instrumentation is usually inadequate. Inaccuracies in the measured reflux ratio can give significant errors in the calculated efficiency, particularly when the column is operating close to minimum reflux. Similar problems can arise because of uncertainty in the vapour–liquid equilibrium data. It should be standard practice to quote measured tray or section efficiencies together with the equilibrium data on which they are based. Regrettably this practice is rarely followed.

Extensive compilations of experimentally determined efficiencies may be found in a number of publications (Perry & Green 1984, Vital *et al.* 1984, Chan & Fair 1984, Kastanek & Standart 1966, 1967).

7.8 Empirical correlations for efficiency

Prior to about 1960, empirical efficiency correlations generally related efficiency to a few key variables, the most important of which was liquid viscosity (Walter & Sherwood 1941, Drickamer & Bradford 1943, O'Connell 1946, Chu *et al.* 1951, Barker & Choudhury 1959, Lockhart & Leggett 1958). Later correlations tended to use the sledgehammer approach and to express efficiency in terms of all the dimensionless groups which could possibly be relevant (English & Van Winkle 1963, Eduljee 1965, MacFarland *et al.* 1972). Surprisingly, both types of correlations are still being recommended (Vital *et al.* 1984).

The best known of the earlier correlations is that of O'Connell (1946) for bubble cap trays which, for distillation, can be expressed as

$$E_0 = 9.06(\mu_L \alpha)^{-0.245} \tag{7.23}$$

where

E_0 = overall column efficiency (%)

μ_L = liquid viscosity (N s m^{-2})

α = relative volatility

The significance of μ_L in eqn. (7.23) is that an increase in μ_L is also usually associated with a decrease in liquid phase diffusivity (Fig. 1.3), so leading to an increase in liquid phase resistance and to a reduction in tray efficiency.

Examination of O'Connell's data shows that there was little justification for including α in the correlation. On the other hand, Lockhart & Leggett (1958) recommended a similar correlation to eqn. (7.23), valid also for absorbers. For absorbers, α was replaced by ten times the equilibrium constant of the key component. Here again this can be interpreted as a fall in efficiency because of an increase in liquid phase resistance caused by an increase in the slope of the equilibrium line.

Typical of the empirical correlations of the second type involving dimensionless groups is that of MacFarland *et al.* (1972)

$$E_{MV} = 7.0\left(\frac{\sigma}{\mu_L u_s}\right)^{0.14}\left(\frac{\mu_L}{\rho_L D_L}\right)^{0.25}\left(\frac{h_w u_h \rho_G}{\mu_L}\right)^{0.08} \tag{7.24}$$

or

$$E_{MV} = 6.8\left(\frac{\sigma}{\mu_L u_s}\right)^{0.115}\left(\frac{\mu_L}{\rho_L D_L}\right)^{0.215}\left(\frac{h_w u_h \rho_G}{\mu_L}\right)^{0.1} \tag{7.25}$$

These equations were based on data from both bubble cap and sieve trays.

Perversely, they indicate that E_{MV} is virtually independent of liquid viscosity.

Generally, empirical correlations can only be expected to give very rough estimates of efficiency. They have a place in preliminary studies and when used to support predictions using other more soundly based methods, such as are outlined in subsequent chapters.

8

Point efficiency

8.1 Basic equations for predicting efficiency for a binary mixture

As discussed in Chapter 7, various definitions of efficiency are possible. The approach used in the remaining chapters is based on Murphree efficiencies since these are now firmly established as the most convenient for practical use.

First, the number of vapour-phase and liquid-phase transfer units, N_G and N_L, are estimated from correlations. Then the following four calculation steps are used, which represent a basic procedure for estimating a section efficiency for a binary or pseudo-binary mixture:

number of overall vapour-phase transfer units N_{OG}

$$\frac{1}{N_{OG}} = \frac{1}{N_G} + \frac{1}{N_L} \tag{8.1}$$

Murphree vapour-phase point efficiency E_{OG}

$$E_{OG} = 1 - \exp(-N_{OG}) \tag{8.2}$$

Murphree vapour-phase tray efficiency E_{MV}

$$E_{MV} = \frac{1}{\lambda}[\exp(\lambda E_{OG}) - 1] \tag{8.3}$$

section efficiency E_0 (eqn. (7.11))

$$E_0 = \frac{\ln[1 + E_{MV}(\lambda - 1)]}{\ln \lambda}$$

In Chapters 8 and 9, the origin and use of eqns. (8.1)–(8.3) is discussed in some detail. It is explained when they break down and the use of alternative equations is suggested for those cases. Correlations required for using the equations are also reviewed. Note that to obtain correlations for N_G and N_L it is necessary to work backwards from measured values of either E_0 or

E_{MV}. Consequently, correlations for N_G and N_L are only as good as the accuracy of the equations used in their determination.

8.2 Transfer unit definitions

The following equations define the necessary mass transfer coefficients based on the two-resistance theory of mass transfer which is depicted in Fig. 8.1. The molar flux N is given by

$$N = k_G(y_i - y) = k_L(x - x_i) = K_{OG}(y^* - y) \tag{8.4}$$

where y^* is the vapour composition in equilibrium with the bulk liquid composition:

$$y^* = mx + b \tag{8.5}$$

Equilibrium is assumed at the interface:

$$y_i = mx_i + b \tag{8.6}$$

Combining these equations gives the well-known relationship

$$\frac{1}{K_{OG}} = \frac{1}{k_G} + \frac{m}{k_L} \tag{8.7}$$

Bird *et al.* (1960) give more details about the approximations involved in using eqn. (8.4). The topic is also re-examined in Section 8.6.

8.2.1 *Definitions of* N_{OG}, N_G *and* N_L

Fig. 8.2 shows the elemental strips of the two-phase dispersion on a tray used to define transfer units. A steady state mass balance on the vapour in the shaded element gives

$$G'W\,dz'\,dy = K_{OG}(y^* - y)aW\,dz'\,dh \tag{8.8}$$

In writing eqn. (8.8), it is assumed that the vapour is in plug flow and that there is no horizontal vapour mixing. The number of differential transfer

Fig. 8.1. Concentration profiles of more volatile component in a binary mixture.

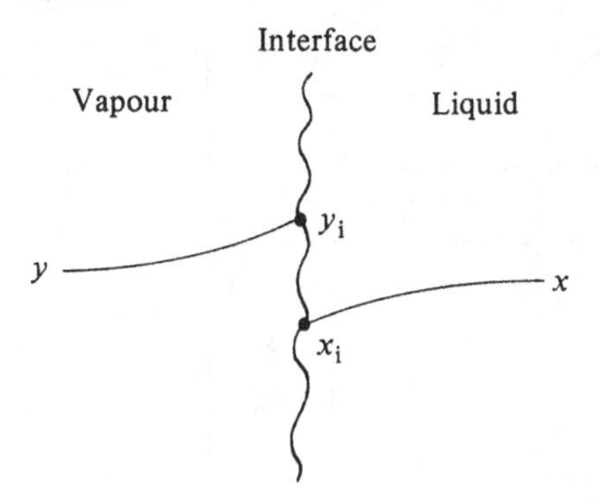

units is defined as the ratio of the vapour concentration change over the differential element to the driving force causing the change. So that from eqn. (8.8), the number of overall vapour-phase transfer units N_{OG} is given by

$$N_{OG} = \int dN_{OG} = \int_{y_{n-1}}^{y_n} \frac{dy}{y^* - y} \tag{8.9}$$

and

$$N_{OG} = \int_0^{h_f} \frac{K_{OG} a \, dh}{G'} = \frac{K_{OG} a h_f}{G'} \tag{8.10}$$

The number of vapour-phase transfer units N_G is obtained when there is no liquid-phase resistance, so that $K_{OG} = k_G$ and

$$N_G = \frac{k_G a h_f}{G'} \tag{8.11}$$

The number of liquid-phase transfer units, N_L, is obtained when there is no vapour-phase resistance, so that from eqn. (8.7), $K_{OG} = k_L/m$, and from eqn. (8.10)

$$N_L = \frac{k_L a h_f}{m G'} \tag{8.12}$$

From eqn. (8.7) it follows that:

$$\frac{1}{N_{OG}} = \frac{1}{N_G} + \frac{1}{N_L} \tag{8.1}$$

Note that eqn. (8.1) can be obtained directly from eqn. (8.7) by multiplying by G'/ah_f. A major advantage of defining N_G and N_L in this way is that we can expect correlations for N_G and N_L to be similar in form.

Transfer units may be expressed in terms of the mean vapour residence time, t_G, in the two-phase dispersion as follows:

$$t_G = \frac{h_f}{u_G} = \frac{h_f \varepsilon}{u_s} = \frac{h_f \varepsilon \rho'_G}{G'} \tag{8.13}$$

Fig. 8.2. Elemental strips used to define transfer units.

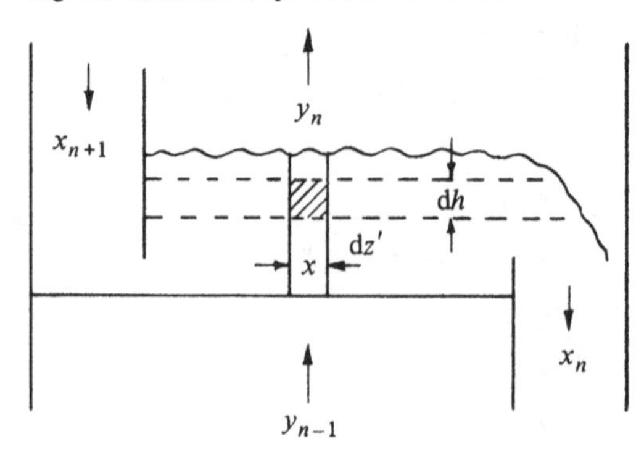

where u_G is the mean vapour velocity through the dispersion and u_s is the superficial vapour velocity. It follows that:

$$N_{OG}=K'_{OG}a't_G \tag{8.14}$$

$$N_G=k'_Ga't_G \tag{8.15}$$

$$N_L=\frac{k'_La't_G}{m}\frac{\rho'_L}{\rho'_G} \tag{8.16}$$

where

$$K'_{OG}=K_{OG}/\rho'_G \tag{8.17}$$

$$k'_G=k_G/\rho'_G \tag{8.18}$$

$$k'_L=k_L/\rho'_L \tag{8.19}$$

$$a'=a/\varepsilon \tag{8.20}$$

and

$$N_{OG}=\frac{(k'_Gah_f)/u_s}{1+\dfrac{mk'_G}{k'_L}\dfrac{\rho'_G}{\rho'_L}} \tag{8.21}$$

8.2.2 N'_L – *an alternative liquid-phase transfer unit*

Traditionally, a different definition of N_L has been used. Assuming that the liquid flows horizontally in plug flow and that there is no vertical liquid mixing, a steady state mass balance on the liquid in the shaded area of Fig. 8.2 gives

$$L\frac{dh}{h_f}\,dx=k_L(x_i-x)aW\,dh\,dz'$$

so that

$$N'_L=\int dN'_L=\int_{x_{n+1}}^{x_n}\frac{dx}{x_i-x}$$

and

$$N'_L=\int_0^Z\frac{k_Lah_fW\,dz'}{L}=\frac{k_Lah_fWZ}{L} \tag{8.22}$$

Substituting into eqn. (8.7) gives

$$\frac{1}{N_{OG}}=\frac{1}{N_G}+\frac{\lambda}{N'_L} \tag{8.23}$$

where

$$\lambda=\frac{mG}{L} \tag{8.24}$$

with $G=G'WZ$. Note that

$$N_L=N'_L/\lambda \tag{8.25}$$

The apparent reason why this rather unrealistic tray model was originally used to define N'_L was that it results in eqn. (8.23) which is the same equation used for packed columns. However, the phases are in countercurrent flow in a packed column and it is an artificial contrivance to force the same equation to apply to the cross flow situation on trays.

N'_L is often expressed in terms of the mean liquid residence time on the tray as follows:

$$\text{horizontal liquid velocity } u_L = \frac{L}{\rho'_L \alpha h_f W}$$

mean liquid residence time $t_L = Z/u_L$, so that from eqn. (8.22)

$$N'_L = k'_L \bar{a} t_L \qquad (8.26)$$

where

$$\bar{a} = \frac{a}{\alpha}$$

Based on eqn. (8.26), many correlations for N'_L have been expressed in terms of t_L. However, N'_L can equally well be correlated in terms of t_G, since

$$t_L = \frac{G\rho'_L \alpha t_G}{L\rho'_G \varepsilon} \qquad (8.27)$$

The use of N'_L rather than N_L as the definition of the number of liquid transfer units has led to considerable confusion. Ellis & Hardwick (1969) and Jeromin *et al.* (1969) argued, incorrectly, that it is possible to alter the relative vapour- and liquid-phase resistances to mass transfer by altering the ratio G/L, as is indicated by a superficial examination of eqn. (8.23). But changing G/L by changing L, for example, will also change N'_L, as shown by eqn. (8.22), with the result that the relative phase resistances remain essentially unchanged. Another objection to using N'_L is that eqn. (8.26) gives the impression that N'_L, and hence N_{OG} and the point efficiency, depend on the mean liquid residence time on the tray. This is obviously incorrect. It appears to be so because only one part of the effect of L is considered in eqn. (8.26). In reality, the combined effects of L in both eqns. (8.23) and (8.26) must be considered together. Furthermore, it should now be clear that the implication given in the *Bubble Tray Design Manual* (AIChE 1958) that some systems can be vapour phase controlled simply because $t_G \ll t_L$ is also incorrect. A final criticism of the use of N'_L is that it makes it more difficult to develop correlations for the number of transfer units. By correlating data in terms of N_G and N_L, the similarities between vapour and liquid phase controlled mass transfer can be exploited. There is considerable potential for improving tray efficiency prediction methods by recorrelating existing data in terms of N_L and N_G rather than N'_L and N_G.

8.3 Relationship between E_{OG} and N_{OG}

In order to define the point efficiency in a practically useful way, it is necessary to assume that the liquid is completely mixed in the vertical direction. This actually is a very good assumption which is easily verified by injecting dye into the liquid at a point on the tray. Standart *et al.* (1979) have given a model in which this assumption was not made, but the complexity of the model precludes its use for design purposes.

Referring to Fig. 8.2, the Murphree vapour-phase point efficiency is

$$E_{OG} = \frac{y_n - y_{n-1}}{y^* - y_{n-1}} \tag{8.28}$$

where $y^* = mx + b$ and x is the local liquid concentration at the point on the tray.

For plug flow of vapour and with y^* constant, integration of eqn. (8.9) gives

$$N_{OG} = -\ln\left(\frac{y^* - y_n}{y^* - y_{n-1}}\right)$$

so that

$$E_{OG} = 1 - \exp(-N_{OG}) \tag{8.29}$$

Alternatively, if the vapour is completely mixed vertically, a mass balance gives

$$G'W\,dz'(y_n - y_{n-1}) = K_{OG}(y^* - y_n)aW\,dz'h_f$$

from which it follows that:

$$E_{OG} = \frac{N_{OG}}{1 + N_{OG}} \tag{8.30}$$

So far as is known, the extent of vapour backmixing on a distillation tray has not been measured. It is probably of minor importance in the spray and froth regimes but in the emulsion regime it could be significant.

Of much more importance than vapour backmixing is the way in which vapour passes through the froth. As shown in Section 2.4, the vapour travels as voids having a range of sizes and velocities. A model for point efficiency involving a polydisperse bubble size distribution is derived below.

Let f_i be the fraction of the total volumetric vapour flow carried through the froth by bubbles of equivalent spherical diameter d_{bi} and rise velocity u_{bi}. Referring to Fig. 8.3, the residence time of these bubbles in the volume element of froth shown is $\Delta h/u_{bi}$. The volume flow rate of bubbles of subgroup i is $u_s A f_i$.

If the flow is suddenly stopped, a flow rate out of the element of $u_s A f_i$ will occur for a time $\Delta h/u_{bi}$. It follows that the holdup fraction of bubbles of subgroup i in the element is

$$\varepsilon_i = \frac{(u_s A f_i \,\Delta h)/u_{bi}}{A\,\Delta h} = \frac{u_s f_i}{u_{bi}} \tag{8.31}$$

The total gas holdup fraction in the froth is therefore

$$\varepsilon = u_s \sum_{i=1}^{n} \frac{f_i}{u_{bi}} \tag{8.32}$$

where n is the number of bubble subgroups and

$$\sum_{i=1}^{n} f_i = 1$$

The number of bubbles of subgroup i in volume $A\,\Delta h$ is $6\varepsilon_i A\,\Delta h/\pi d_{bi}^3$ and interfacial area contributed by subgroup $i = 6\varepsilon_i A\,\Delta h \pi d_{bi}^2/\pi d_{bi}^3$. Interfacial area of bubble subgroup i per unit volume of froth is

$$a_i = \frac{6\varepsilon_i}{d_{bi}} \tag{8.33}$$

Total interfacial area per unit volume of froth from eqns. (8.31) and (8.33) is

$$a = 6u_s \sum_{i=1}^{n} \frac{f_i}{d_{bi} u_{bi}} \tag{8.34}$$

We assume plug flow of each bubble through the froth with complete mixing of liquid vertically. A steady state mass balance for bubbles of subgroup i over a differential height dh of froth gives

$$u_s A f_i \rho'_G \,\mathrm{d}y = K_{OGi}(y^* - y) a_i A \,\mathrm{d}h \tag{8.35}$$

Substituting and rearranging

$$\frac{\mathrm{d}y}{y^* - y} = \frac{6K_{OGi}\,\mathrm{d}h}{d_{bi} u_{bi} \rho'_G} \tag{8.36}$$

Integrating over the total height of the froth and rearranging gives the point efficiency or fractional approach to equilibrium for bubbles of subgroup i:

Fig. 8.3. Idealised froth containing a range of bubble sizes.

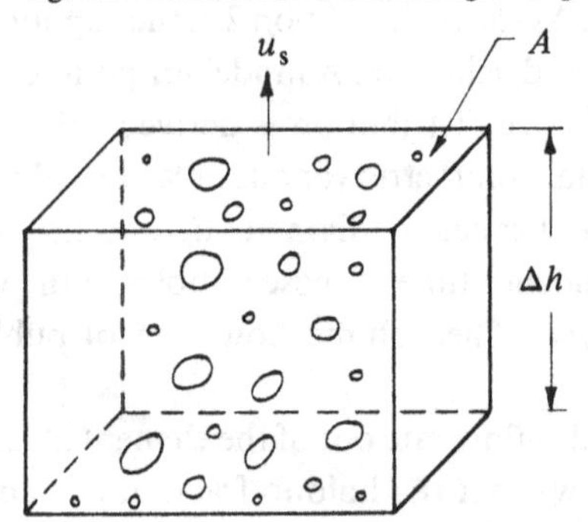

$$E_{OGi}=1-\exp\left(-\frac{6K_{OGi}h_f}{d_{bi}u_{bi}\rho'_G}\right) \tag{8.37}$$

For the vapour as a whole, the point efficiency is

$$E_{OG}=\sum_{i=1}^{n} f_i E_{OGi}=1-\sum_{i=1}^{n} f_i \exp\left(-\frac{6K_{OGi}h_f}{d_{bi}u_{bi}\rho'_G}\right) \tag{8.38}$$

Ashley & Haselden (1972) argued that the distribution of bubble sizes in a froth may be described, at least approximately, in terms of two dominant bubble sizes, d_{b1} and d_{b2}, having rise velocities u_{b1} and u_{b2}. They defined ε'_1 as the local volume fraction of small bubbles, i.e. the volume of small bubbles per unit volume of that part of the froth which contains only small bubbles. By simple mass balances

$$u_{b2}=\frac{u_s(1-\varepsilon'_1)}{\varepsilon-\varepsilon'_1}-\frac{u_{b1}\varepsilon'_1(1-\varepsilon)}{\varepsilon-\varepsilon'_1} \tag{8.39}$$

and

$$f_1=\frac{u_{b1}\varepsilon'_1(1-\varepsilon)}{u_s(1-\varepsilon'_1)}, \quad f_2=1-f_1 \tag{8.40}$$

Basing N_{OG} on the total interfacial area we have

$$N_{OG}=\frac{K_{OG}ah_f}{\rho'_G u_s} \tag{8.41}$$

For simplicity, we assume $K_{OG1}=K_{OG2}=K_{OG}$ so that, for a two-bubble size dispersion, eqn. (8.38) becomes

$$E_{OG}=1-f_1\exp\left(\frac{-6N_{OG}u_s}{ad_{b1}u_{b1}}\right)-f_2\exp\left(\frac{-6N_{OG}u_s}{ad_{b2}u_{b2}}\right) \tag{8.42}$$

Although it is difficult to assign accurate values to u_{b1} and ε'_1, u_{b1} is probably larger than the terminal rise velocity of isolated bubbles because of liquid circulation. A value of $u_{b1}=0.6\ \mathrm{m\ s^{-1}}$ and two alternate values of ε'_1 of 0.5 and 0.7 have been used to construct Fig. 8.4, using eqn. (8.42) together with eqns. (8.34), (8.39) and (8.40).

Eqn. (8.42) is a third relationship between E_{OG} and N_{OG} which is compared with eqns. (8.29) and (8.30) in Fig. 8.4. Very significant differences in E_{OG} result from the same value of N_{OG} depending on the model used. In the spray regime, eqn. (8.29) probably holds fairly well because of the likelihood of a uniform vapour residence time in the dispersion. But in the froth regime, to which eqn. (8.42) applies, the binary bubble-size model predicts a much smaller point efficiency than the two other models when N_{OG} is greater than about 0.1 ($E_{OG}>7$–10%).

The reason for this is that when the extent of mass transfer is high (corresponding to large N_{OG}), the small bubbles reach equilibrium with the liquid before they leave the froth. Consequently, the effective interfacial area is lower than the actual interfacial area. Another way of looking at it is that when E_{OG} is larger than about 10%, its value is dominated by the behaviour of the large bubbles. Most of the vapour passes through the froth as large bubbles. Although the small bubbles are more numerous and apparently give a large interfacial area for mass transfer, they contribute only marginally to the overall vapour concentration change. This means that the total interfacial area is irrelevant in determining the point efficiency. What is important is the approach to equilibrium of the vapour carried by the large bubbles. This is illustrated in Fig. 8.5 where E_{OG} and interfacial area have been calculated as a function of the diameter of the small bubbles using eqn. (8.38). It shows that, although the interfacial area changes by an order of magnitude, E_{OG} remains constant.

The model leading to eqn. (8.38) is obviously oversimplified. It neglects enhanced mass transfer during bubble formation, for example, and there are indications that significant mass transfer takes place in this region (Aerov *et al.* 1970, Lockett *et al.* 1979). But at least the model does show that eqn. (8.29), which is almost universally used to relate N_{OG} and E_{OG}, is grossly in error at all but low values of E_{OG}. The implications of this are taken up again in Section 8.5.3. Furthermore, the topic of non-uniform residence times of vapour passing through a froth has a close analogy with that of non-uniform residence times of liquid crossing a tray – both lower

Fig. 8.4. Relationship between E_{OG} and N_{OG} using different models. $\varepsilon = 0.78$, $u_{b1} = 0.6 \text{ m s}^{-1}$, $u_s = 1.68 \text{ m s}^{-1}$, $d_{b1} = 5 \times 10^{-3}$ m, $d_{b2} = 50 \times 10^{-3}$ m.

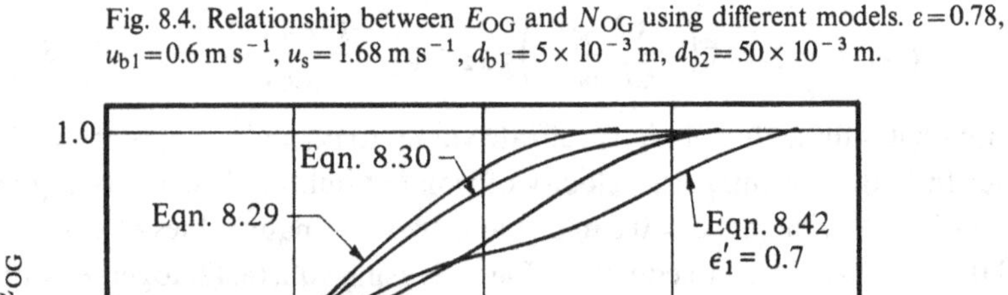

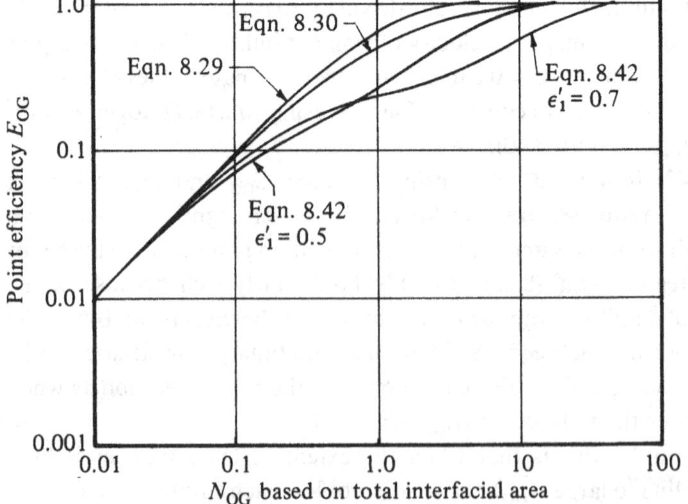

efficiency. The latter is dealt with in Section 9.13. Other workers who have reached similar conclusions regarding polydisperse bubble-size distributions in froths are Kaltenbacher (1982) and Hofer (1983).

8.4 Estimation of E_{OG} using an Oldershaw column

One approach to estimating the tray efficiency in full-size columns is to measure the efficiency obtained in an Oldershaw column containing the same system. Some designers, wary of correlations and equations, prefer the sense of security this brings, particularly with unfamiliar systems. Fair *et al.* (1983) found that the technique worked well provided that both the Oldershaw and full-size columns were operated at the same fractional approach to flooding. Because of its small diameter, an Oldershaw column gives an estimate of the point efficiency. For this to be the same as on a full-size tray, the values of $(K_{OG}ah_f)/(\rho'_G u_s)$ must be the same in both cases based, perhaps incorrectly, on eqn. (8.29). Fair *et al.* explained this apparently fortuitous equality as follows: at the same approach to flooding

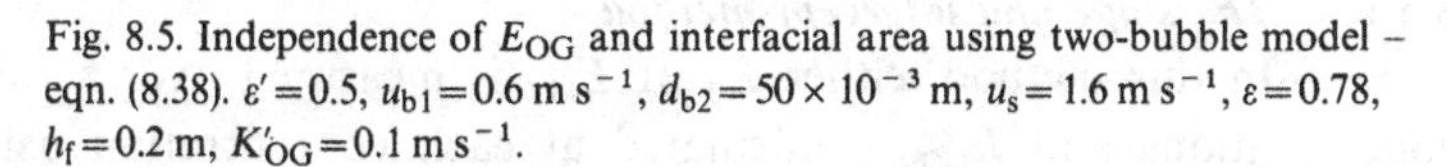

$$\frac{h_{f1}}{T_{s1}} \approx \frac{h_{f2}}{T_{s2}}$$

where 1 refers to the Oldershaw column and 2 to the full-size column. h_f is the froth height and T_s is the tray spacing. This implies that flooding is jet

Fig. 8.5. Independence of E_{OG} and interfacial area using two-bubble model – eqn. (8.38). $\varepsilon' = 0.5$, $u_{b1} = 0.6$ m s^{-1}, $d_{b2} = 50 \times 10^{-3}$ m, $u_s = 1.6$ m s^{-1}, $\varepsilon = 0.78$, $h_f = 0.2$ m, $K'_{OG} = 0.1$ m s^{-1}.

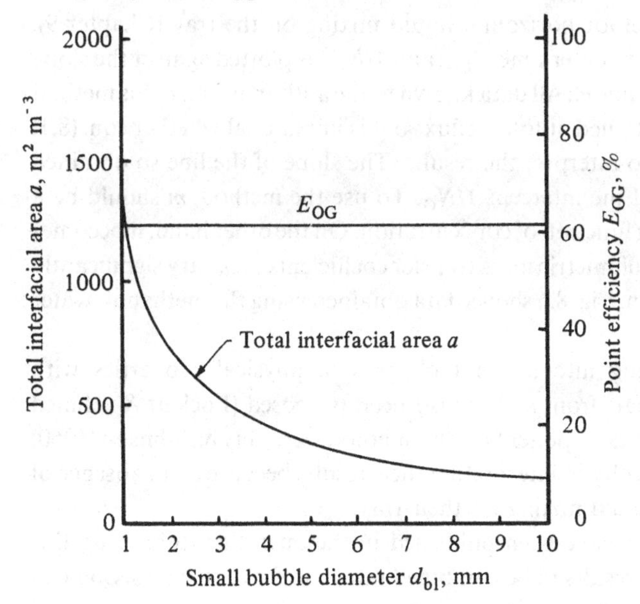

flooding in both cases. In Fair *et al.*'s study $T_{s1} = 25$ mm, $T_{s2} = 610$ mm and the ratio of the flooding velocities was about 3. It follows that for equal E_{OG}:

$$(K_{OG}a)_1 = 8(K_{OG}a)_2$$

The enhanced value of $(K_{OG}a)$ in an Oldershaw column can perhaps be attributed to the much smaller holes used and also to the lower superficial vapour velocities. Both tend to give a finer dispersion with a minimum of vapour bypassing as large bubbles.

The usefulness of an Oldershaw column for estimating E_{OG} appears to depend on a fortunate combination of circumstances. Particular care should be used when, in the full-size column, the potential flooding mechanism is likely to be by downcomer backup, the tray spacing is other than 610 mm and the weir height is different from the value used by Fair *et al.* of 50 mm.

8.5 Estimation of E_{OG} from correlations

Volumetric mass transfer coefficients (ka) are usually either correlated directly or in combination with h_f and u_s as N_G, N_L or N'_L. The various approaches which have been proposed to measure and correlate these quantities are discussed in the following sections.

8.5.1 *The slope and intercept method*

In this method, either E_0 or E_{MV} is measured over a range of concentrations and E_{OG} is calculated at each concentration using an appropriate model for horizontal liquid mixing on the tray (Chapter 9). Eqn. (8.29) is used to determine N_{OG}, and $1/N_{OG}$ is plotted against the slope of the equilibrium line m. All data known to the author in which this method was used were obtained at total reflux so it is immaterial whether eqn. (8.1) or (8.23) is used to interpret the results. The slope of the line so obtained gives $G'/k_L a h_f$ and the intercept $1/N_G$. To use the method, m should be a reasonably strong function of concentration. On the other hand, it becomes inaccurate if the volumetric mass transfer coefficients or h_f vary significantly with concentration. Fig. 8.6 shows data obtained using the methanol–water system.

A way of taking into account changes in physical properties with concentration, apart from m, has also been proposed (Lockett & Ahmed 1983). Two of the first exponents of this method were Hay & Johnson (1960) but they had difficulty in interpreting their results because of an absence of information on liquid mixing on their tray.

Insufficient data have been published in the open literature using this procedure for the results to be combined into a useful correlation. Some of the results which have been obtained are shown in Table 8.1, and they indicate that distillation systems have significant liquid-phase resistance.

Zuiderweg's correlation. Based primarily on FRI data, which was analysed using the slope and intercept method, and also by using a semi-theoretical expression for k'_G, Zuiderweg (1982) proposed the following equations:

$$k'_G = \frac{0.13}{\rho_G} - \frac{0.065}{\rho_G^2} \quad (1.0 < \rho_G < 80) \tag{8.43}$$

$$k'_L = 2.6 \times 10^{-5} \mu_L^{-0.25} \tag{8.44}$$

or

$$k'_L = 0.024(D_L)^{0.25} \tag{8.45}$$

Fig. 8.6. $1/N_{OG}$ vs. slope of equilibrium line m for methanol–water. (Lockett & Ahmed 1983.)

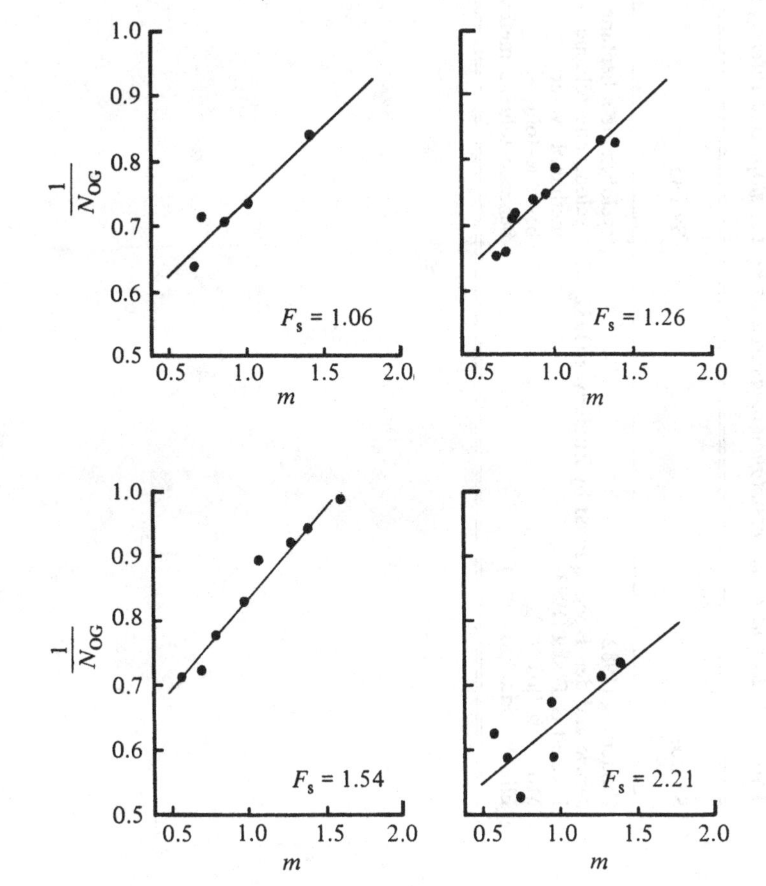

Table 8.1. *Liquid phase resistance measured using the slope and intercept method for distillation*

Source	System	% Liquid phase resistance ($100\,N_{OG}/N_L$)
Zuiderweg (1982)	cyclohexane-*n* heptane	50
Jancic and den Hoed, quoted by Zuiderweg (1982)	cyclohexane–toluene	37
Lockett & Plaka (1983)	methanol–water	30
Moens & Bos (1972)	benzene–toluene	≈50
Ellis & Biddulph (1967)	benzene–toluene–methyl cyclohexane	2–32

Spray regime:

$$ah_f = \frac{40}{\phi^{0.3}}\left(\frac{u_s^2 \rho_G h_{cl} FP}{\sigma}\right)^{0.37} \tag{8.46}$$

Mixed-froth/emulsion flow regimes:

$$ah_f = \frac{43}{\phi^{0.3}}\left(\frac{u_s^2 \rho_G h_{cl} FP}{\sigma}\right)^{0.53} \tag{8.47}$$

Eqn. (3.26) is used to determine h_{cl} in eqns. (8.46) and (8.47). These equations can be used to determine N_{OG} from eqn. (8.21).

The inclusion of surface tension in the above equations is open to question. When using the FRI data, it is difficult to decide whether surface tension or some other physical property is important because, for the systems used, physical properties tend to change in unison, see Section 1.3.1. When these equations are applied to aqueous systems, which tend to be surface tension positive, they predict low point efficiencies, about 40%, whereas experimental efficiencies typically are higher at about 80%. Zuiderweg (1983) attempted to explain this discrepancy by suggesting that the Marangoni effect can cause froth stabilisation for surface tension positive systems, such as methanol–water, and can increase the point efficiency even at vapour velocities bordering on the transition to the spray regime. This is hard to accept. Its test awaits experimental data from surface tension negative systems having a high surface tension, e.g. ethylene glycol–water, which should exhibit a low point efficiency if Zuiderweg's interpretation is correct.

8.5.2 *Correlations based on absorption, stripping and humidification*

A great deal of data has been obtained from absorption, stripping or humidification experiments, where, depending on the system used, m is either very low or very high. From eqn. (8.7), the resistance to mass transfer can be confined in this way to either the gas or liquid phase, respectively, and so either $k_G a$ or $k_L a$ can be determined.

AIChE correlations. Using the procedure outlined above, Gerster and co-workers (Gerster *et al.* 1958, AIChE 1958, Smith 1963) developed the following correlations in an AIChE sponsored programme:

$$N_G = \left(0.776 + 4.57\, h_w - 0.238\, F_s + 104.8\, \frac{Q_L}{W}\right) Sc_G^{-0.5} \tag{8.48}$$

$$Sc_G = \frac{\mu_G}{\rho_G D_G} \tag{8.49}$$

and

$$N_L' = 1.97 \times 10^4 (D_L)^{0.5} (0.40\, F_s + 0.17) t_L \tag{8.50}$$

where

$$t_L = \frac{h_{cl} Z W}{Q_L} \tag{8.51}$$

These equations are used with eqns. (8.23) and (8.29) to give E_{OG}. In eqn. (8.51), h_{cl} is given by eqn. (3.27) with the parameters given by Gerster (quoted in Smith 1963) for sieve trays shown in Table 3.3. Eqn. (8.48) was developed for both bubble cap and sieve trays, whereas eqn. (8.50) holds for sieve trays only. Note that neither correlation involves surface tension. The importance of the Schmidt number in eqn. (8.48) is open to doubt. For example, in very careful experiments, Mehta & Sharma (1966) found that N_G depends on D_G but not on Sc_G.

Table 8.2 lists some other correlations which have been proposed which are similar to the AIChE correlations.

Table 8.2. *Transfer unit correlations based on absorption, stripping and humidification*

Harris (1965), sieve trays

$$N_G = (0.3 + 15 t_G) Sc_G^{-0.5}$$

$$N'_L = [5 + 10 t_L (1 + 0.17(0.82 F_s - 1)(39.3 h_w + 2))] Sc_L^{-0.5}$$

Asano & Fujita (1966), sieve trays

$$N_G = 5.85\, Sc_G^{-0.5} \left(\frac{u_h d_h \rho_G}{\mu_G} \right)^{-0.25} P$$

$$N'_L = 460 \left(\frac{n d_h}{D} \right) Sc_L^{-0.5} \left(\frac{h_{cl}}{d_h} \right)^{-0.5} P \qquad (n = \text{number of holes})$$

where

$$P = \left(\frac{u_h^2 \rho_G}{d_h g \rho_L} \right)^{0.17} \left(\frac{h_{cl}}{d_h} \right)^{0.15} \left(\frac{d_h^2 g \rho_L}{\sigma} \right)^{0.1}$$

Jeromin *et al.* (1969), sieve trays

$$N_G = Sc_G^{-0.5} \left(1.2 + 4.57\, h_w - 0.238\, F_s + 106 \frac{Q_L}{W} \right) + \Gamma \qquad (\Gamma = \text{'foam factor'})$$

$$N'_L = 2.03 \times 10^4 (1 + \lambda^2)(D_L)^{0.5} (0.175\, F_s + 0.15) t_L$$

Hughmark (1971), bubble cap trays

$$N_G = (0.051 + 0.0105\, F_s) \left(\frac{\rho_L}{F_s} \right)^{0.5}, \quad \text{where } \left(\frac{\rho_L}{F_s} \right) < 1089$$

$$N'_L = \left(-44 + 1.08 \times 10^5 \frac{Q_L}{W} + 127 F_s \right) \frac{A_b}{Q_L} (D_L)^{0.5}$$

where

$$\frac{Q_L}{W} < 0.007, \quad 1.1 < F_s < 2.4$$

8.5.3 *Percentage liquid-phase resistance in distillation*

The AIChE correlations give typical values of N_G and N'_L obtained from separate absorption or stripping studies in which the resistance is confined to either the gas or liquid phase. Zuiderweg's correlations give typical values of N_G and N'_L obtained from distillation data using the slope and intercept method.

Values of N_G and N'_L calculated using the two different approaches are shown in Fig. 8.7. (At total reflux with $m=1$, $\lambda=1$, so $N_L=N'_L$.) The percentage liquid-phase resistance defined as

$$\left(\frac{100}{N_L}\right)\bigg/\left(\frac{1}{N_{OG}}\right)$$

is shown in Fig. 8.8. The major difference between the two approaches is that the AIChE method gives larger N'_L values which leads to a smaller percentage liquid-phase resistance and to a higher predicted point efficiency – Fig. 8.9. The results of two other prediction methods, Chan & Fair (1984) and Stichlmair (1978), are also shown in Figs. 8.8 and 8.9, but are discussed later.

The reason why Zuiderweg's method gives a higher liquid-phase resistance can be understood by referring to Fig. 8.4. Using the slope and intercept method with actual distillation data, N_{OG} is determined from E_{OG} using eqn. (8.29) and the values of N_{OG} obtained are used to give N_G and N'_L. The use of eqn. (8.29) involves a gross approximation to the real situation, which is more realistically represented by eqn. (8.42), except perhaps in the

Fig. 8.7. Comparison of N'_L and N_G predictions. $D=2.0$ m, $W/D=0.75$, single-pass sieve tray, $T_s=0.61$ m, $h_w=0.05$ m, $d_h=0.0127$ m, $p=0.038$ m, $\phi=0.1$, total reflux, $m=1$, 80% flood. Using eqns. (5.1)–(5.3), physical properties from Fig. 1.3.

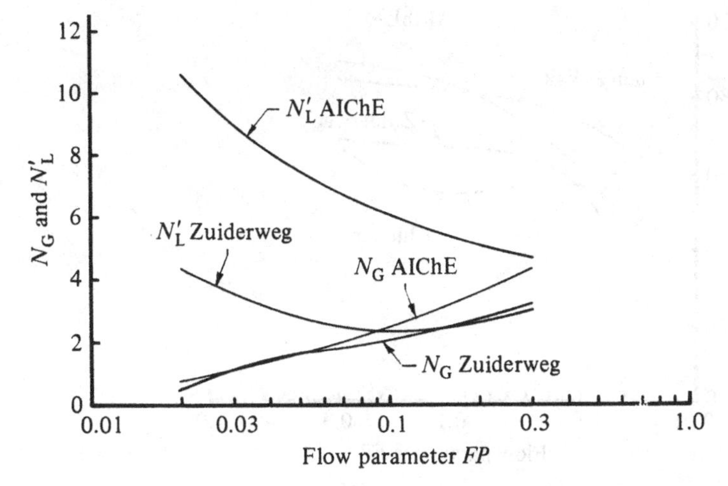

spray regime. But it does at least lead to the same error in both N_G and N'_L when using the slope and intercept method. Consequently, the percentage liquid-phase resistance, which involves the ratio of N_G and N'_L, tends to be estimated correctly. However, the absolute values of N_G and N'_L depend strongly on the model used to relate E_{OG} and N_{OG}. For design purposes this causes no problems providing the same model (eqn. (8.29)) is used to determine N_G and N'_L from E_{OG} for correlation purposes and also to predict E_{OG} from N_G and N'_L for design.

Fig. 8.8. Predicted percentage liquid phase resistance – conditions as for Fig. 8.7.

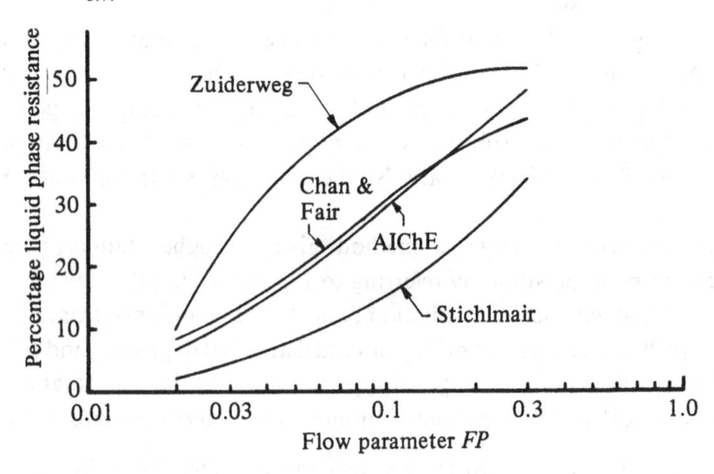

Fig. 8.9. Predicted point efficiency – conditions as for Fig. 8.7.

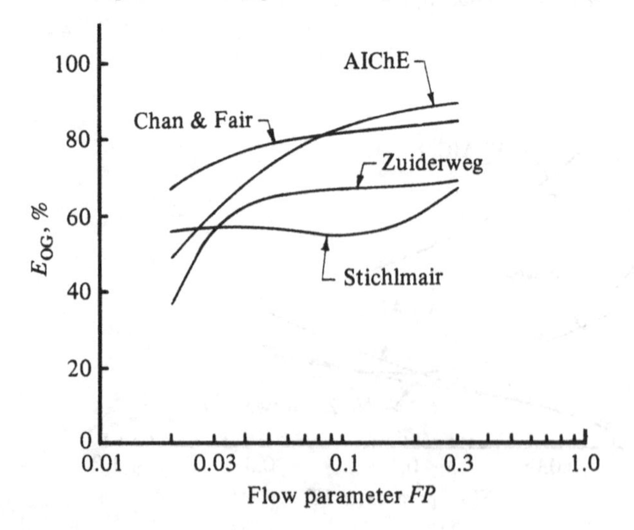

Unfortunately, the situation is different using the AIChE approach and other approaches of the same type. To determine N'_L, an absorption or stripping system is used where the gas is relatively insoluble in the liquid (high m). It follows that $N_{OG} = N'_L \cdot L/mG$ and N_{OG} and E_{OG} are very small. For example, E_{OG} is typically about 0.01 for absorption of CO_2 into water from air for which $m = 1420$ at 20°C. The procedure used is to measure E_{ML} (Section 7.2), to calculate E_{MV} (eqn. (7.10)), to determine E_{OG} using a liquid mixing model and finally to use eqn. (8.29) to give N_{OG} and hence N'_L. Fig. 8.4 shows that for small values of E_{OG}, the value of N_{OG} obtained is independent of the model used to relate E_{OG} and N_{OG}. Consequently, the values of N'_L obtained in this way should be reasonably accurate.

To measure N_G, absorption of a soluble gas or humidification of air is used for which m is either close to or equal to zero. The mass transfer resistance is confined to the gas phase so that $N_{OG} = N_G$. Now the values of N_{OG} and E_{OG} are much higher and typically E_{OG} is between 0.6 and 0.8. Unfortunately, in this range, the value of N_{OG} determined from the measured value of E_{OG} depends very strongly on the model used – Fig. 8.4. Use of eqn. (8.29) gives values of N_{OG} (and N_G) some 4–20 times smaller than are obtained from the more realistic binary bubble-size model – eqn. (8.42). It follows that the consistent use of eqn. (8.29) leads to an underestimate of N_G but not of N'_L. This tends to underestimate the percentage liquid-phase resistance. Consequently in a design situation, when N_G and N'_L are combined to give N_{OG} and eqn. (8.29) is used to determine E_{OG}, the method has a bias towards overprediction of point efficiency. Further detailed analysis of this problem has been given by Lockett & Plaka (1983).

8.5.4 *Chan & Fair's correlation*

Insofar as they based their correlation on an extensive data bank of distillation data, Chan & Fair's (1984) correlation can be considered as a hybrid lying between the two approaches represented by the Zuiderweg and the AIChE correlations. Chan & Fair obtained values of E_{OG} from distillation data and then used eqn. (8.29) to determine N_{OG}. The AIChE correlation, eqn. (8.50), was used to give N'_L and values of N_G were calculated from eqn. (8.23). This procedure is open to the same criticism as applies to the AIChE procedure, namely that it tends to underestimate the percentage liquid-phase resistance. Offsetting this is the redeeming feature that the correlation is based on actual distillation data. The final correlation obtained is

$$N'_L = 1.97 \times 10^4 (D_L)^{0.5} (0.40\, F_s + 0.17) t_L \tag{8.52}$$

with

$$t_L = h_{cl} Z W / Q_L \tag{8.53}$$

$$N_G = \frac{1000(D_G)^{0.5}(10.3(FF) - 8.67(FF)^2)t_G}{h_{cl}^{0.5}} \tag{8.54}$$

with

$$t_G = (1 - \alpha_e)h_{cl}/(\alpha_e u_s) \tag{8.55}$$

In the above, (FF) is the fractional approach to flooding, $(FF) = u_s/(u_s$ at flooding), and h_{cl} and α_e are calculated using eqns. (3.23) and (3.24). Values of E_{OG} and percentage liquid-phase resistance calculated using the above equations, together with eqns. (8.23) and (8.29), are included in Figs. 8.8 and 8.9. Fig. 8.8 shows that the method predicts almost exactly the same percentage liquid-phase resistance as the AIChE method. This is not unexpected since the same correlation for N'_L is used in both methods. It also shares with the AIChE method a tendency to predict high values of E_{OG}. Note that the inclusion of (FF) in the N_G correlation, eqn. (8.54), implies that efficiency depends on tray spacing for fixed vapour and liquid loads. This hardly seems reasonable and the presence of (FF) serves as a reminder about the empiricism inherent in the correlation.

8.5.5 Estimation of E_{OG} from individual values of k' and a

In this section we deal with point efficiency prediction methods, apart from that of Zuiderweg, in which attempts have been made to estimate k' and a individually rather than using correlations involving the combined volumetric coefficients $k'_G a$ and $k'_L a$. Proposed methods have appeared regularly over the years (Geddes 1946, Chu *et al.* 1951, West *et al.* 1952, Garner & Porter 1960, Calderbank & Moo-Young 1960, Andrew 1961, de Goederen 1965, Fane & Sawistowski 1969), but none of them has gained any practical acceptance. Two possible exceptions are the methods proposed by Stichlmair (1978) and by Neuburg & Chuang (1982).

Stichlmair's model for point efficiency. Stichlmair's model combines eqns. (8.21) and (8.29) with the following equations for a and k'. When $F_s/F_{smax} \leqslant 0.7$

$$a = a_B - \left(\frac{F_s/F_{smax}}{0.7}\right)^2 (a_B^* - a_T^*) \tag{8.56}$$

where

$$a_B = 6\left[\frac{(\rho_L - \rho_G)g}{6\sigma}\right]^{0.5}\left(\frac{F_s}{F_{smax}}\right)^{0.28}$$

$$a_B^* = 6\left[\frac{(\rho_L - \rho_G)g}{6\sigma}\right]^{0.5}(0.7)^{0.28}$$

$$a_T^* = \frac{(0.7\,F_{smax})^2}{2\sigma\phi^2}(1 - 0.7^{0.28})$$

When $F_s/F_{smax} > 0.7$

$$a = \frac{F_s^2}{2\sigma\phi^2}\left[1 - (F_s/F_{smax})^{0.28}\right] \tag{8.57}$$

Mass transfer coefficients are calculated from

$$k'_G = 2\left(\frac{D_G u_s}{\pi h_f \varepsilon}\right)^{0.5} \tag{8.58}$$

and

$$k'_L = 2\left(\frac{D_L u_s}{\pi h_f \varepsilon}\right)^{0.5} \tag{8.59}$$

F_{smax} and ε (or $1-\alpha$) are given by the correlations attributed to Stichlmair in Table 3.2, h_{cl} is given by eqn. (3.20) and $h_f = h_{cl}/(1-\varepsilon)$. In the limiting case, when $F_s \to 0$, Stichlmair calculated the Sauter mean bubble diameter d_b from eqn. (2.2) corresponding to bubble formation under surface-tension dominated conditions. In the spray regime ($F_s > 0.7\,F_{smax}$ according to Stichlmair) the mean drop size d_p was determined from

$$d_p = \frac{12\sigma}{\rho_G}\left(\frac{\phi}{u_s}\right)^2$$

Interfacial area a was then calculated for the two limiting values of F_s from

$$a = \frac{6\varepsilon}{d_b} \quad (\text{for } F_s \to 0) \quad \text{and} \quad a = \frac{6(1-\varepsilon)}{d_p} \quad \left(\text{for } \frac{F_s}{F_{smax}} > 0.7\right)$$

Interpolation results in eqns. (8.56) and (8.57) for interfacial area a over the whole range of F_s.

This approach can be criticised because bubble formation under surface-tension dominated conditions has little relevance to the situation on sieve trays at normal vapour flow rates, where bubble formation size does not depend on surface tension. In froths, large bubbles or vapour voids are formed by bubble coalescence and their size is not dependent on the size of the bubbles which issue from the holes. The vapour voids serve to transport the large volumetric vapour flow through the froth by virtue of their high rise velocity. It follows that the strong influence of surface tension on interfacial area in the froth regime, as indicated by eqn. (8.56), cannot be justified. A further reservation is that interfacial area is based on uniform size bubbles and the consequences of this simplification have been discussed in Section 8.3. These assumptions result in excessively large values of interfacial area using eqn. (8.56). In compensation, Stichlmair's equations for k'_G and k'_L result in overly small values. Eqns. (8.58) and (8.59) are based on the Higbie penetration theory (Higbie 1935), where

$$k'_G = 2\left(\frac{D_G}{\pi\theta}\right)^{0.5} \quad \text{and} \quad k'_L = 2\left(\frac{D_L}{\pi\theta}\right)^{0.5} \tag{8.60}$$

In eqn. (8.60), θ is the contact time of an element at the gas–liquid interface. In Higbie's original work, θ was taken as the time for a bubble to rise through its own height. In contrast, Stichlmair argued that the appropriate time was the time for the vapour to travel through the dispersion on the tray, $\theta = h_f \varepsilon / u_s$. He justified this by arguing that gross liquid circulation in the dispersion results in a negligible slip velocity between vapour and liquid. The argument is speculative and particularly inappropriate for the spray regime. Its result is to base k' on the maximum possible value of θ.

Stichlmair claimed that his model gave good agreement with experimental results, but it gives poor agreement with the predictions of the other correlations shown in Figs. 8.8 and 8.9. One reason for the discrepancy is the difficulty of accurately predicting F_{smax}. For the conditions used in Figs. 8.8 and 8.9, F_s is close to F_{smax}. As a consequence, spray-regime operation is incorrectly predicted over most of the range of conditions shown when using Stichlmair's criteria of $F_s/F_{smax} > 0.7$ for spray.

Neuburg and Chuang's model for point efficiency. Although Neuburg & Chuang (1982) developed their model specifically for predicting froth-regime point efficiencies on trays used in the Girdler–Sulfide process for the production of heavy water, it can be easily extended to distillation in general. For prediction of the mass-transfer coefficients they used Higbie's equation (eqn. (8.60)) with $\theta = d_b/u_b$ and $u_b = u_s/\varepsilon$, where d_b and u_b are the Sauter mean bubble size and the corresponding bubble rise velocity. Since $a = 6\varepsilon/d_b$, substituting into eqn. (8.21) and rearranging gives

$$K'_{OG}a = 12\left(\frac{D_L u_s \varepsilon}{\pi d_b^3}\right)^{0.5}\left[\left(\frac{D_L}{D_G}\right)^{0.5} + \frac{m\rho'_G}{E'\rho'_L}\right]^{-1} \tag{8.61}$$

E' is an enhancement factor to take into account chemical reaction and $E' = 1$ for distillation. Eqn. (8.61) is used with eqn. (8.29) to give E_{OG}. Neuburg & Chuang concluded that no satisfactory correlation exists for d_b so they determined it by back calculation from measured values of E_{OG} in pilot-scale experiments. The values of d_b so obtained were used to predict E_{OG} in full-size columns. A very similar procedure was also adopted by Calderbank & Pereira (1979) in which they used measured bubble sizes and velocities from the 'Calderbank probe' – Section 2.4.

8.5.6 *Estimation of k'_G, k'_L and a using mass transfer with chemical reaction*

By absorbing a solute from a gas stream into a liquid wherein it undergoes a chemical reaction, it is possible to determine individual values of k'_G, k'_L and a. The technique has been well documented (Sharma & Gupta 1967, Sharma *et al.* 1969, Sharma & Danckwerts 1970, Danckwerts & Sharma 1966, Danckwerts 1970) and is not elaborated here. Some of the results obtained are shown in Table 8.3.

In spite of considerable effort in measuring these individual parameters for a wide variety of tray types and operating conditions, the results have not yet been used to any significant extent for the design of trays in commercial distillation columns. More surprisingly perhaps, in the authors experience they have also been little used for the design of trays in acid gas absorption and stripping columns. Some possible explanations for this unsatisfactory state of affairs are offered below:

(i) The technique for determining interfacial area is largely only applicable to aqueous solutions containing electrolytes. Bubble coalescence is known to be inhibited in such solutions and the interfacial areas obtained may be larger than those found in hydrocarbon systems. The importance of bubble coalescence appears to depend on the tray type. Sharma *et al.* (1969), for example, found that for sieve trays interfacial area increased strongly with ionic strength, whereas the same effect was not found for bubble cap trays. A possible explanation is that vapour bypassing was greater using bubble cap trays.

(i) There is poor agreement between interfacial areas measured by different workers. Stichlmair (1978) has given a comparison of most of the data available.

(iii) As argued in Section 8.3, a knowledge of only the interfacial area is insufficient to predict point efficiency except when the point efficiency is very low. The techniques used in its measurement often involved only a small change in the solute gas concentration over the height of the froth (Porter *et al.* 1966, Sharma *et al.* 1969). Under these circumstances the smallest bubbles would not have reached equilibrium and the value of interfacial area obtained was probably close to the true interfacial area of the froth. However, most workers neglected to report the solute gas concentration change. In some studies it is possible that the smallest bubbles may have reached equilibrium, and this could be a contributing factor to the wide range of interfacial areas reported.

8.6 Point efficiency and heat transfer

8.6.1 *Introduction*

An omission from the point efficiency prediction methods considered so far is their neglect of sensible heat transfer from the vapour to

Table 8.3. *Reported values of a, k'_L and k'_G using mass transfer with chemical reaction*

Source	a ($m^2\ m^{-3}$)	$k'_L \times 10^4$ ($m\ s^{-1}$)	k'_G ($m\ s^{-1}$)
Bubble cap trays			
Andrew (1961)[a]	$324\ u_s^{0.5} h_f^{-0.17}$	$3.6 \times 10^4\ u_s^{0.25} h_f^{-0.5} (D_L)^{0.5}$	$2.26\ u_s^{0.25} h_f^{-0.5} (D_G)^{0.5}$
	(313)	$(7.6 \times 10^4 (D_L)^{0.5})$	$(4.8 (D_G)^{0.5})$
Porter *et al.* (1966)	140–224	4	
Sharma *et al.* (1969)	$244\ u_s^{0.5} h_f^{-0.17}$	$4.1 \times 10^4\ u_s^{0.25} h_f^{-0.5} (D_L)^{0.5}$	$3.6\ u_s^{0.25} h_f^{-0.5} (D_G)^{0.5}$
	(235)	$(8.4 \times 10^4 (D_L)^{0.5})$	$(7.6 (D_G)^{0.5})$
McNeil (1970)	200	2.8	
McLachlan & Danckwerts (1972)	$8.3 \times 10^3\ u_s^{1.6} h_{cl}^{0.45}$		$0.88\ h_{cl}^{-0.65} (D_G)^{0.5}$
	(643)		$(7.1 (D_G)^{0.5})$
Takeuchi *et al.* (1977)	200–600	28–56	
Sieve trays			
Eben & Pigford (1965)	900–2500		
Smith & Wills (1966)	≈65		
Pohorecki (1968)	240		
Sharma *et al.* (1969)	100–250		
Pasiuk-Bronikowska (1969)	200	17–27	
Bartholomai *et al.* (1972)	185–470	6.5–12.3	
Pohorecki (1976)	250	7–11	
Thomas & Haq (1976)	150–240	5–6.3	
Hofer (1983)		froth 2–3	
		spray 6–10	
Dual flow sieve trays			
Sharma & Gupta (1967)	200–400	1.5–4.5	
Rodionov & Vinter (1967)	≈475	10–40	

[a] In Andrew's correlation, slot submergence taken as h_f

( ) refers to calculated value for $u_s = 0.5\ m\ s^{-1}$, $h_f = 0.16$ m, $h_{cl} = 0.04$ m

the liquid – in distillation the vapour is hotter than the liquid which it contacts. Also, no account has been taken of differences in the molar latent heats of the components. Opinions differ about whether these omissions make any appreciable difference when predicting point efficiency. It has been argued that point efficiency sometimes passes through a maximum with concentration, particularly for systems where heat transfer could be significant, e.g. methanol–water. Ordinary methods do not predict such a maximum, whereas it is claimed that allowing for heat transfer does do so (Ruckenstein & Smigelschi 1967). The argument is not conclusive because froth stabilisation from the Marangoni effect also varies with concentration and could also be responsible for a maximum in the point efficiency. Furthermore, the observed maxima occurred at very low vapour velocities – a particularly favourable situation for Marangoni effects to play a part.

Fig. 8.10 shows composition and temperature profiles and the direction of the molar fluxes N and the heat fluxes q near the interface during distillation of a binary mixture, where component 1 is the more volatile component.

8.6.2 *Empirical equations*

Several empirical approaches have been proposed in which it is assumed that heat transfer causes additional vaporisation of the liquid which then augments mass transfer by diffusion.

Fig. 8.10. Fluxes, composition and temperature profiles near the phase boundary.

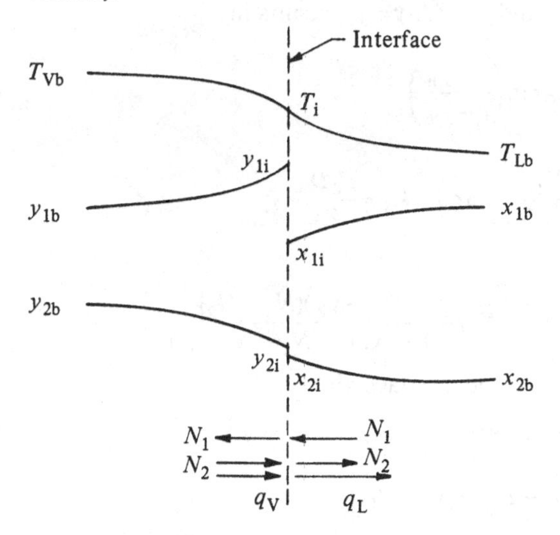

Ruckenstein & Smigelschi (1965) proposed

$$N_1 = k_G(y_{1i} - y_{1b}) + \frac{q_L}{l_1}(y_1^* - x_{1i}) \tag{8.62}$$

where y_1^* is in equilibrium with x_{1b}. Zhavoronkov *et al.* (1979) criticised the above equation and suggested

$$N_1 = k_G(y_{1i} - y_{1b}) + \frac{q_L}{l_1}(y_{1i} - x_{1i}) \tag{8.63}$$

Both equations were claimed by their authors to agree well with experimental data. However, to do so, Ruckenstein & Smigelschi had to assume that about 90% of the resistance to mass transfer lay in the liquid phase, which is very unlikely – see Section 8.5.3. The equations can both be faulted because they simply add the effects of heat transfer to mass transfer. A better approach is to allow for interactions between heat and mass transfer such as is described in the following section.

8.6.3 *Film theory model*

The theory which follows is an adapted version of the treatments given by Sawistowski *et al.* (1964), Todd & Van Winkle (1972) and Pohjola (1973).

For component 1, in the vapour phase, Fick's law of diffusion gives

$$N_1 = y(N_1 - N_2) - \rho'_G D_G \frac{dy}{dz} \tag{8.64}$$

Separating the variables and integrating over the vapour film between the limits $z=0$, $y=y_{1i}$ and $z=Z_F$, $y=y_{1b}$ results in

$$N_1 = -k_G r \ln\left(\frac{r - y_{1i}}{r - y_{1b}}\right) \tag{8.65}$$

where

$$r = \frac{N_1}{N_1 - N_2} \quad \text{and} \quad k_G = \frac{\rho'_G D_G}{Z_F}$$

Rearranging gives

$$N_1 = y_{1b}(N_1 - N_2) + \frac{(y_{1i} - y_{1b})(N_1 - N_2)}{1 - \exp[-(N_1 - N_2)/k_G]} \tag{8.66}$$

An energy balance at the interface yields

$$l_1 N_1 - l_2 N_2 = q \tag{8.67}$$

where

$$q = h_G(T_{Vb} - T_i) - h_L(T_i - T_{Lb}) \tag{8.68}$$

A mean latent heat $\bar{l}$ can be defined such that

$$(N_1 - N_2) = \frac{q}{\bar{l}} \tag{8.69}$$

where

$$\bar{l} = \frac{l_1 N_1 - l_2 N_2}{N_1 - N_2} \quad (l_1 \neq l_2) \tag{8.70}$$

When $l_1 = l_2$, obviously $\bar{l} = l_1 = l_2$. Substituting in eqn. (8.66) and assuming that $q/(\bar{l}k_G)$ is small (Norman 1960) gives

$$N_1 = y_{1b}\frac{q}{\bar{l}} + k_G(y_{1i} - y_{1b}) \tag{8.71}$$

This equation is the same as that quoted by Sawistowski *et al.* (1964). Combining eqns. (8.69) and (8.71) and rearranging leads to the equation used by Pohjola (1973):

$$N_1 = rk_G(y_{1i} - y_{1b})/(r - y_{1b}) \tag{8.72}$$

Starting with a rearranged form of eqn. (8.65), i.e.

$$N_1 = k_G r \ln\left(\frac{r - y_{1b}}{r - y_{1i}}\right) \tag{8.73}$$

leads to the equation used by Todd & Van Winkle (1972):

$$N_1 = y_{1i}(N_1 - N_2) + k_G(y_{1i} - y_{1b}) \tag{8.74}$$

(Note that eqns. (8.71), (8.72) and (8.74) are equivalent.) For component 2,

$$N_2 = y_{2i}(N_1 - N_2) + k_G(y_{2i} - y_{2b}) \tag{8.75}$$

Application of the above theory leads to a similar set of equivalent equations for the liquid phase. That corresponding to eqn. (8.74), for example, is

$$N_1 = x_{1i}(N_1 - N_2) + k_L(x_{1b} - x_{1i}) \tag{8.76}$$

Equilibrium at the interface can be represented by

$$y_{1i} = mx_{1i} + b \tag{8.77}$$

From eqns. (8.69), (8.74), (8.76) and (8.77)

$$x_{1i} = \frac{\dfrac{k_L x_{1b}}{k_G} + y_{1b} - b\left(\dfrac{q}{\bar{l}k_G} + 1\right)}{\dfrac{q(m-1)}{\bar{l}k_G} + \dfrac{k_L}{k_G} + m} \tag{8.78}$$

Given correlations for the heat and mass transfer coefficients, Todd & Van Winkle have indicated how similar equations to those developed above can be used to determine the flux of each component across the interface. It is necessary to assume that each phase is saturated. An iterative solution is involved with initial assumptions that

$$q=0 \quad \text{and} \quad \bar{l}=\frac{l_1+l_2}{2}$$

Eqn. (8.78) is then solved iteratively for x_{1i} (note that m and b depend on x_{1i}), and y_{1i} is obtained from eqn. (8.77). N_1 and N_2 can be determined from eqns. (8.74) and (8.75). Better estimates of q and $\bar{l}$ can then be obtained from eqns. (8.68) and (8.70), respectively, and the procedure repeated to convergence. If an estimate of the interfacial area is available, the point efficiency can finally be determined from the mass flux of each component. The calculation procedure described above is the only one known to the author which comes even close to approaching a practical design method for distillation point efficiency prediction involving heat transfer.

If plug flow of vapour is assumed through the froth, however (an assumption which was not made by Todd & Van Winkle), the procedure becomes excessively complicated because the variation of y_{1b} through the froth then has to be taken into account. Furthermore, correlations for the heat and mass transfer coefficients and interfacial area are not reliable. The question then naturally arises whether there is anything to be gained by using such a complex procedure and whether the accuracy obtained using the much simpler prediction methods of Section 8.5 is acceptable.

Pohjola (1973) has argued that, even using the equations developed above, the final equation for prediction of point efficiency which results does not involve inter-phase heat transfer or the molar latent heats. His argument proceeds as follows: for plug flow of vapour, total and component balances give

$$\mathrm{d}G'=(N_1-N_2)a\,\mathrm{d}h$$

and

$$\mathrm{d}(G'y_{1b})=N_1 a\,\mathrm{d}h \tag{8.79}$$

Expanding and rearranging results in

$$\frac{\mathrm{d}G'}{G'}=\frac{\mathrm{d}y_{1b}}{r-y_{1b}}$$

Hence

$$\mathrm{d}(G'y_{1b})=G'\,\mathrm{d}y_{1b}\left(1+\frac{y_{1b}\,\mathrm{d}G'}{G'\,\mathrm{d}y_{1b}}\right)=\left(\frac{r}{r-y_{1b}}\right)G'\,\mathrm{d}y_{1b} \tag{8.80}$$

From eqns. (8.72), (8.79) and (8.80)

$$\frac{r}{r-y_{1b}}G'\,\mathrm{d}y_{1b}=\frac{rk_G(y_{1i}-y_{1b})a\,\mathrm{d}h}{r-y_{1b}}$$

Cancelling gives

$$G' \, dy_{1b} = k_G(y_{1i} - y_{1b})a \, dh \tag{8.81}$$

Remarkably, eqn. (8.81) is equivalent to the simple eqn. (8.8) which took no account of heat transfer or inequality of the latent heats. This implies that the point efficiency does not depend on these effects.

8.6.4 *Other approaches*

One shortcoming of the above theories is that they are based on the film theory, i.e. $k_G \propto D_G$, whereas in practice it has been found, at least in absorption and stripping, that $k_G \propto D_G^{0.5}$ is more realistic. Ruckenstein (1970) has given an analysis based on the penetration theory, but the results are very complex and difficult to apply.

Recently Sandall and co-workers (Kayihan *et al.* 1975, 1977, Honorat & Sandall 1978, Sandall & Dribika 1979), taking an approach similar to that of Todd & Van Winkle, but without assuming that the bulk phases were saturated, also concluded that heat-transfer effects are negligible in distillation. Any reassurance which one can draw from this conclusion is offset somewhat by a further conclusion drawn by Sandall that liquid-phase resistance to mass transfer is always negligible during distillation. The mathematical arguments leading to the latter conclusion are not entirely convincing, however, and the notion has not gained any acceptance.

A more pragmatic approach for deciding on the importance of heat transfer during distillation was taken by Norman (1960). He estimated the maximum relative rates of heat and mass transfer for methanol–water distillation and concluded that the rate of sensible heat transfer was not more than about 5% of the heat transferred as latent heat. This implies that sensible heat transfer has a negligible effect.

It may be concluded that, for design purposes, sensible heat transfer and unequal latent heats may be ignored during distillation. All point efficiency prediction methods used in practice of which the author is aware do indeed make this assumption.

8.6.5 *Inter-tray heat transfer*

When there is a considerable difference in temperature between the vapour below a tray and the liquid on the tray and in the downcomer, heat transfer can take place through the downcomer wall and through the tray deck. For very small columns, this leads to an increase in the efficiency of the tray–downcomer combination as demonstrated by Ellis & Shelton (1960) using a 0.1 m-diameter column. Lockett & Ahmed (1983) gave a more-detailed analysis of the problem for a 0.6 m-diameter column

Table 8.4. *Effect of individual variables on* E_{OG}

	Froth regime			Spray regime			
Variable	E_{OG} increases	E_{OG} falls	No effect	E_{OG} increases	E_{OG} falls	No effect	Sources
Increasing surface tension			0		0		Sawistowski (1978)
		0			0		Zuiderweg (1982); see eqns. (8.46), (8.47)
		0				0	Zuiderweg (1984)
Increasing surface tension gradient σ^{+} systems	0						Fane & Sawistowski (1969)
Increasing surface tension gradient σ^{-} systems				0			Sawistowski (1978)
Increasing weir height	0					0	Hai *et al.* (1977)
Increasing vapour velocity		0		0			Hai *et al.* (1977)
							Lockett *et al.* (1979)
Increasing hole diameter		0					Lockett *et al.* (1979)
				0			Hai *et al.* (1977)
Increasing hole area		0					Kreis & Raab (1979)

distilling methanol–water, and, for this size column at least, the increase in efficiency was found to be negligible. Here again, it seems that in practical situations heat-transfer effects can safely be disregarded.

8.7 Use of point efficiency models in design

It should now be apparent that, at best, point efficiencies can only be predicted to about $\pm 10\%$ using the models available. However, the situation is not as bad as it seems. Only in the rarest of cases is a designer called upon to predict efficiency without any experience whatsoever of a similar distillation system. A designer who finds himself in such an unenviable position is well advised to seek help from someone who has such experience. Once a reference point is available for the measured efficiency of a comparable operating column, then point efficiency models are an invaluable tool for extrapolating to different conditions. Typical different conditions might involve changes in pressure, temperature, vapour velocity, liquid load, weir height, etc. To complement model predictions, Table 8.4 summarises the reported effects of some individual variables on point efficiency. The only really confused area is the effect of surface tension. As discussed in Section 2.8.2.3, enhanced interfacial area is obtained for σ^+ systems in the froth regime and for σ^- systems in spray. Fell & Pinczewski (1977) have suggested taking advantage of this to maximise efficiency by designing σ^+ and σ^- systems to operate in the froth and spray regimes, respectively.

9

Relationship between point efficiency and tray efficiency

9.1 Introduction

In general, the Murphree tray efficiency E_{MV} is greater than the Murphree point efficiency E_{OG}. In favourable circumstances, E_{MV} can exceed 100%. This arises because the liquid is never perfectly mixed on the tray in the horizontal direction and, as mass transfer proceeds, a concentration gradient is established in the liquid as it moves across the tray. This in turn causes a concentration gradient in the vapour leaving the tray. Fig. 9.1 illustrates the concentration profile of the more volatile component in the exit vapour for a binary mixture. The inlet vapour is assumed perfectly mixed. We have

$$E_{MV} = \frac{\bar{y}_n - \bar{y}_{n-1}}{y_n^* - \bar{y}_{n-1}} \quad \text{where } y_n^* = m\bar{x}_n + b$$

Fig. 9.1. Concentration profile of more volatile component in the vapour leaving a tray.

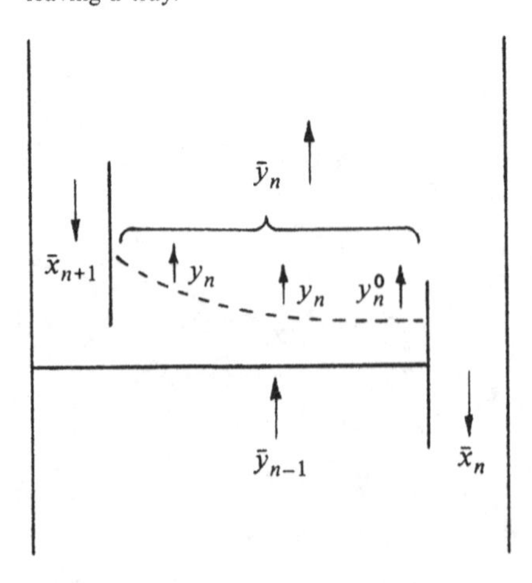

and locally at the liquid exit

$$E_{OG} = \frac{y_n^0 - \bar{y}_{n-1}}{y_n^* - \bar{y}_{n-1}}$$

Since $\bar{y}_n > y_n^0$, it follows that $E_{MV} > E_{OG}$.

9.2 Lewis's three cases

Lewis (1936) was the first to determine quantitative relationships between E_{MV} and E_{OG} for the special situation of no horizontal liquid mixing on the tray (plug flow). The three alternative cases considered by Lewis and shown in Fig. 9.2 are:

Lewis's case 1: vapour is completely mixed between trays. The direction of liquid flow on successive trays is immaterial.

Lewis's case 2: vapour is unmixed between trays. Liquid flows in the same direction on successive trays.

Lewis's case 3: vapour is unmixed between trays. Liquid flows in alternate directions on successive trays.

The equations derived by Lewis are:

case 1

$$E_{MV} = \frac{\exp(\lambda E_{OG}) - 1}{\lambda} \tag{9.1}$$

case 2

$$E_{MV} = \frac{\alpha - 1}{\lambda - 1} \tag{9.2}$$

Fig. 9.2. Representation of Lewis's three cases.

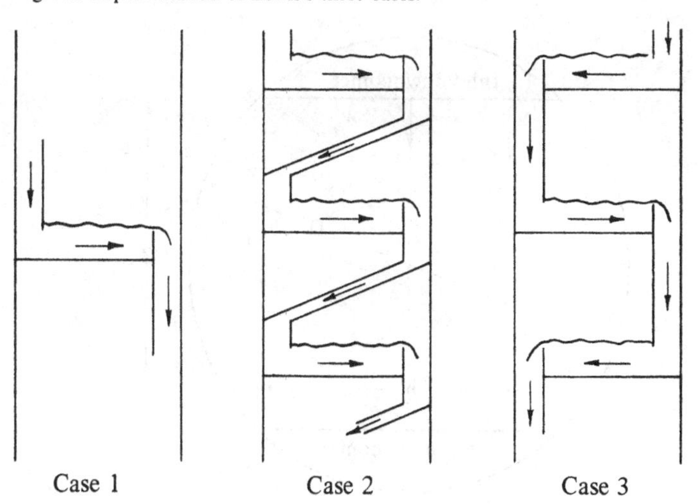

where

$$\lambda = \left(\frac{1}{E_{OG}} + \frac{1}{\alpha - 1}\right)\ln \alpha \tag{9.3}$$

Here α is the concentration similarity ratio.

case 3

$$E_{MV} = \frac{\alpha - 1}{\lambda - 1}$$

where, if $\alpha < 1$,

$$\lambda = \left[\frac{\alpha^2 - (1 - E_{OG})^2}{E_{OG}^2(1 - \alpha^2)}\right]^{0.5} \cos^{-1}\left[1 - \frac{(1-\alpha)(\alpha - 1 + E_{OG})}{\alpha(2 - E_{OG})}\right] \tag{9.4}$$

and if $\alpha > 1$

$$\lambda = \left[\frac{\alpha^2 - (1 - E_{OG})^2}{E_{OG}^2(\alpha^2 - 1)}\right]^{0.5} \cosh^{-1}\left[1 + \frac{(\alpha - 1)(\alpha - 1 + E_{OG})}{\alpha(2 - E_{OG})}\right] \tag{9.5}$$

For cases 2 and 3, α is determined from eqns. (9.3)–(9.5) as appropriate and used in eqn. (9.2) to give E_{MV}. Values obtained from these equations are given as part of the tables in Appendix B for $Pe = 1000$. Note that case 2 gives the highest tray efficiency and case 3 the lowest. Vapour mixing reduces tray efficiency for case 2 and improves it for case 3 – bringing them both closer to case 1.

9.3 A general equation for local liquid concentration

Lewis's equations give the maximum tray efficiency which can be achieved. In reality the tray efficiency falls short of the maximum value

Fig. 9.3. Coordinate system.

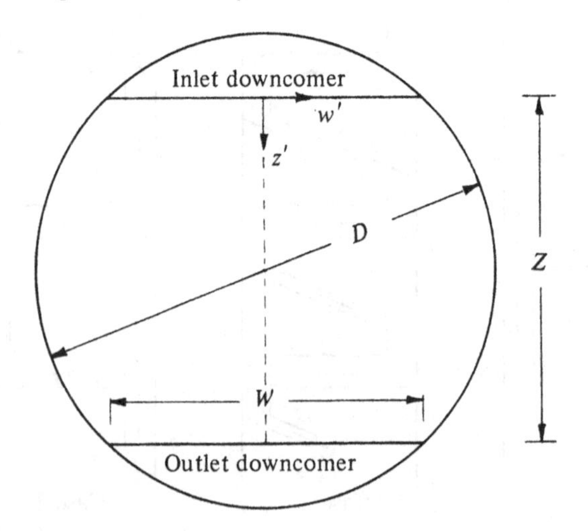

because of:

(1) liquid mixing,
(2) vapour mixing for case 2,
(3) non-uniform liquid flow across the tray,
(4) non-uniform flow of vapour through the tray,
(5) entrainment and weeping.

Mathematical models have been developed for all of the above, either singly or in combination. Some economy of discussion can be achieved by developing a differential equation which, on simplification, is applicable to all five of the situations listed above.

Using the coordinate system of Fig. 9.3, the flows of a particular component into and out of an element of froth or spray located at coordinates (w', z') are shown in Fig. 9.4. Liquid mixing, caused by the vapour passing through it, is represented by an eddy diffusion coefficient De assumed equal in all horizontal directions. At steady state for tray n, the

Fig. 9.4. Mass balance over dispersion on incremental area of tray n, allowing for liquid flow, vapour flow, liquid mixing, entrainment and weeping.

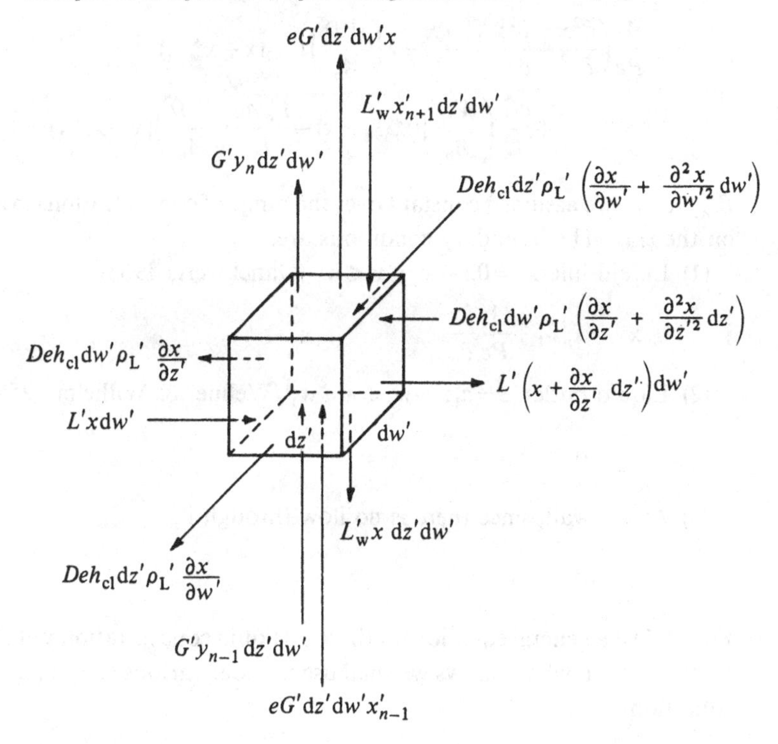

liquid concentration x at point (w', z') is given by

$$De\, h_{cl}\rho'_L\left(\frac{\partial^2 x}{\partial z'^2}+\frac{\partial^2 x}{\partial w'^2}\right)-L'\frac{\partial x}{\partial z'}+G'(y_{n-1}-y_n)$$
$$+eG'(x'_{n-1}-x)-L'_w(x-x'_{n+1})=0 \quad (9.6)$$

These terms represent transfer by mixing of liquid caused by the vapour, bulk liquid flow, transfer to the vapour, entrainment and weeping, respectively. It is assumed that bulk liquid flow takes place in the z' direction only. With the following substitutions:

$$z=z'/D, \quad w=w'/D, \quad L'=L/W,$$
$$z_1=Z/D, \quad w_1=W/2D, \quad G'=G/A_b$$
$$y_n-y_{n-1}=mE_{OG}(x-x^*_{en-1}), \quad \text{where } y_{n-1}=mx^*_{en-1}+b$$
$$Pe=\frac{LD}{Wh_{cl}\rho'_L De}$$

and

$$\lambda=\frac{mG}{L}$$

eqn. (9.6) becomes

$$\frac{1}{Pe}\left(\frac{\partial^2 x}{\partial z^2}+\frac{\partial^2 x}{\partial w^2}\right)-\frac{\partial x}{\partial z}-\lambda\left(\frac{WD}{A_b}\right)E_{OG}(x-x^*_{en-1})$$
$$+\frac{eG}{L}\left(\frac{WD}{A_b}\right)(x'_{n-1}-x)-\frac{L'_w A_b}{L}\left(\frac{WD}{A_b}\right)(x-x'_{n+1})=0 \quad (9.7)$$

E_{OG} and m are assumed constant over the range of concentrations involved on the tray. The boundary conditions are:

(1) Liquid inlet, $z=0$, $-w_1 \leqslant w \leqslant w_1$ (Danckwerts 1953)

$$x_+=\bar{x}_{n+1}+\frac{1}{Pe}\frac{\partial x}{\partial z} \quad (9.8)$$

(2) Liquid outlet, $z=z_1$, $-w_1 \leqslant w \leqslant w_1$ (Wehner & Wilhelm 1956)

$$\frac{\partial x}{\partial z}=0 \quad (9.9)$$

(3) At the wall, since there is no flow through it,

$$\frac{\partial x}{\partial s}=0 \quad (9.10)$$

Eqn. (9.7) is a general equation for the local liquid concentration x at a point on the tray. In what follows we shall use it under various sets of particular conditions.

9.4 Simple backmixing model for Lewis's case 1

For a rectangular tray of width W and length Z, we replace D by Z in the above equations and assume no entrainment or weeping. Only changes of concentration in the z direction are involved and the one-dimensional version of eqn. (9.7) is

$$\frac{1}{Pe}\frac{\mathrm{d}^2x}{\mathrm{d}^2z}-\frac{\mathrm{d}x}{\mathrm{d}z}-\lambda E_{\mathrm{OG}}(x-x^*_{en-1})=0 \tag{9.11}$$

Convenient boundary conditions are at $z=1$, $\mathrm{d}x/\mathrm{d}z=0$ and $x=\bar{x}_n$. The solution to eqn. (9.11) (Gerster *et al.* 1958) for completely mixed entering vapour is

$$\frac{x-x^*_{en-1}}{\bar{x}_n-x^*_{en-1}}=\frac{\exp[(\eta+Pe)(z-1)]}{1+[(\eta+Pe)/\eta]}+\frac{\exp[\eta(1-z)]}{1+[\eta/(\eta+Pe)]} \tag{9.12}$$

with

$$\eta=\frac{Pe}{2}\left[\left(1+\frac{4\lambda E_{\mathrm{OG}}}{Pe}\right)^{0.5}-1\right] \tag{9.13}$$

Eqn. (9.12) gives the liquid concentration as a function of z. Now

$$\bar{y}_n-\bar{y}_{n-1}=mE_{\mathrm{OG}}\int_0^1(x-x^*_{en-1})\,\mathrm{d}z$$

and from eqn. (7.2)

$$\bar{y}_n-\bar{y}_{n-1}=E_{\mathrm{MV}}(y^*_n-\bar{y}_{n-1})=E_{\mathrm{MV}}m(\bar{x}_n-x^*_{en-1})$$

so that

$$\frac{E_{\mathrm{MV}}}{E_{\mathrm{OG}}}=\int_0^1\left(\frac{x-x^*_{en-1}}{\bar{x}_n-x^*_{en-1}}\right)\mathrm{d}z \tag{9.14}$$

Substitution of eqn. (9.12) into eqn. (9.14) gives on integration

$$\frac{E_{\mathrm{MV}}}{E_{\mathrm{OG}}}=\frac{1-\exp[-(\eta+Pe)]}{(\eta+Pe)\{1+[(\eta+Pe)/\eta]\}}+\frac{\exp\eta-1}{\eta\{1+[\eta/(\eta+Pe)]\}} \tag{9.15}$$

Eqn. (9.15) is the relationship between E_{MV} and E_{OG} for Lewis's case 1 for any degree of liquid mixing. When $Pe=\infty$ it simplifies to eqn. (9.1). When the liquid is completely mixed, $Pe=0$, it simplifies to $E_{\mathrm{MV}}=E_{\mathrm{OG}}$. Both are interesting exercises for a rainy afternoon.

9.5 Simple backmixing model for Lewis's cases 2 and 3

Analytical solutions equivalent to eqn. (9.15) for Lewis's cases 2 and 3 with partial liquid mixing have not been obtained. However, numerical solutions have been given by Diener (1967). Alternatively a numerical solution can be obtained by tray-to-tray calculation starting at the bottom of the column (Lockett *et al.* 1973).

Using the nomenclature of Fig. 9.5, the solution of eqn. (9.11) with boundary conditions given by eqns. (9.8) and (9.9) requires values of x^*_{en-1} and $\bar{x}_{n+1}$ to be specified. In general, after tray-to-tray calculations to tray n, the values of x^*_{en-1} are available from calculations for tray $n-1$. For $\bar{x}_{n+1}$, the procedure is to assume a value and solve eqn. (9.11) numerically for x over tray n. The assumed value of $\bar{x}_{n+1}$ is then adjusted and the calculations repeated until the calculated outlet liquid concentration from tray n, $\bar{x}_n$, is equal to the concentration of liquid entering tray $n-1$ used in the calculations for tray $n-1$. From the definition of point efficiency

$$x^*_{en} = x^*_{en-1} + E_{OG}(x - x^*_{en-1}) \tag{9.16}$$

where x refers to the local liquid concentration on tray n. Values of x^*_{en} for use in calculations for tray $n+1$ are obtained from eqn. (9.16). In this way liquid concentrations can be determined for each tray up the column. Corresponding Murphree tray efficiencies can be obtained for each tray by integrating the calculated concentration profiles, i.e. from eqns. (9.14) and (9.16):

$$(E_{MV})_n = \frac{\int_0^1 (x^*_{en} - x^*_{en-1})\,dz}{\bar{x}_n - \int_0^1 x^*_{en-1}\,dz} \tag{9.17}$$

To start the calculations, liquid and vapour leaving the reboiler are taken to be in equilibrium and arbitrary values of x_R and $\bar{x}_2$ are assumed. The calculated efficiencies are independent of these arbitrary assumptions.

Fig. 9.5. Nomenclature for tray-to-tray calculation from the bottom of the column.

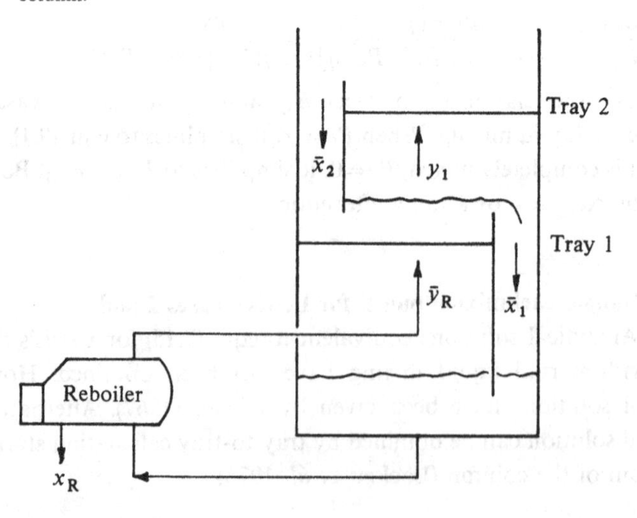

Because the vapour entering the bottom tray is taken as completely mixed, the bottom tray corresponds to Lewis's case 1. Subsequently the vapour is taken as unmixed, and after about four trays the tray efficiency becomes constant and corresponds to Lewis's case 3. Calculations for Lewis's case 2 may be made simply by reversing the vapour concentration profile between each tray to represent liquid flow on successive trays in the same direction.

The results of these calculations for a range of conditions are included in Appendix B. Fig. 9.6 shows the typical variation of E_{MV} with liquid Peclét number.

A useful approximation can be used to calculate E_{MV} for any values of Pe, λ and E_{OG} for Lewis's cases 2 and 3, which avoids a numerical solution to the equations. It is based on the similarity of the variation of E_{MV} with Pe for each of the three Lewis cases – examples of which are shown in Fig. 9.6. A ratio θ is defined for each Lewis case and for fixed values of λ, Pe and E_{OG} as

$$\theta = \frac{(E_{MV})_{Pe} - E_{OG}}{(E_{MV})_{Pe=\infty} - E_{OG}}$$

$(E_{MV})_{Pe}$ is the value of E_{MV} for a particular value of Pe, where $0 < Pe < \infty$. $(E_{MV})_{Pe}$ and $(E_{MV})_{Pe=\infty}$ are evaluated at the same values of λ and E_{OG}. Now θ can be calculated exactly for Lewis's case 1 from eqns. (9.1) and (9.15). As an approximation, θ can be taken as the same for each Lewis case when evaluated at the same values of Pe, λ and E_{OG}.

For example, to determine E_{MV} for Lewis's case 3 for particular values of

Fig. 9.6. Variation of E_{MV} with Pe for $E_{OG} = 0.8$.

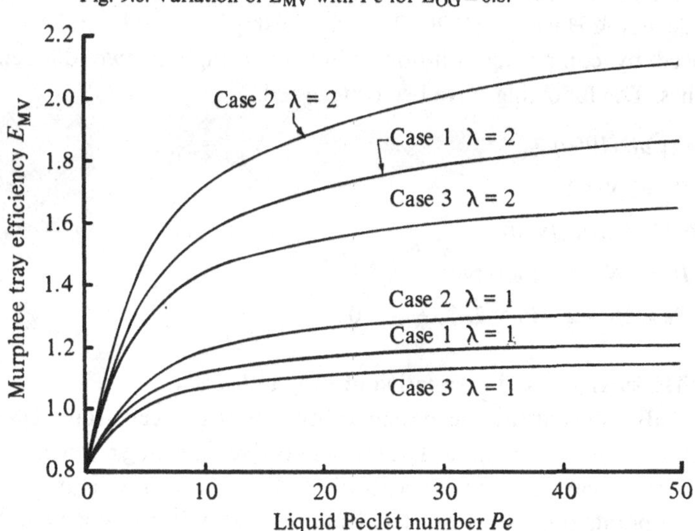

Pe, λ and E_{OG} the following equation is used:

$$E_{MV} = E_{OG} + [(E_{MV})_{\substack{Pe=\infty \\ \text{case 3}}} - E_{OG}] \frac{(E_{MV})_{\substack{Pe \\ \text{case 1}}} - E_{OG}}{(E_{MV})_{\substack{Pe=\infty \\ \text{case 1}}} - E_{OG}} \tag{9.18}$$

$(E_{MV})_{Pe=\infty}$ for case 3 is calculated from eqns. (9.4) or (9.5). Values of E_{MV} obtained from eqn. (9.18) are generally within 1 or 2% of the values shown in Appendix B obtained by numerical methods. The largest error occurs at low values of Pe, where the assumption of no vapour mixing inherent in Lewis's cases 2 and 3 is itself open to question. It follows that eqn. (9.18) is perfectly adequate for all practical purposes.

9.6 Mixed pools model for liquid mixing

An alternative to the eddy diffusion model for liquid mixing is the mixed pools model in which the liquid is envisaged as a series of completely mixed pools extending along the length of the tray. This model has the advantage that it is easy to visualise the situation being modelled. For a series of N pools the tray-to-point efficiency ratio for Lewis's case 1 conditions is (Gautreaux & O'Connell 1955)

$$\frac{E_{MV}}{E_{OG}} = \frac{[1 + (\lambda E_{OG}/N)]^N - 1}{\lambda E_{OG}} \tag{9.19}$$

Obviously, this reduces to eqn. (9.1) in the limit as $N \rightarrow \infty$. Numerical solutions for Lewis's cases 2 and 3 using the mixed pools model have been given by Ashley & Haselden (1970). The relationship between Pe and N can be deduced by comparing solutions obtained using the two different approaches. The following have been suggested:

Williams et al. (1960)

$Pe = 2N - 2$

Ashley & Haselden (1970)

$Pe = 2N - 1$ Pe large

$Pe = 2N - 2$ $Pe > 2$, $\lambda E_{OG} < 0.5$

9.7 Measurement and correlation of eddy diffusivity

Eddy diffusivity can be measured either by an unsteady state (USS) or by a steady state (SS) method. The former involves injecting a tracer into the liquid and measuring the concentration variation with time at either one or two points downstream (Mecklenburgh 1974). In the steady state method, the tracer concentration is measured at a series of points upstream of the injection point (Barker & Self 1962).

Reliable results are easier to obtain using the steady state method and eddy diffusivity is simpler to deduce from the data. Consequently, most workers have preferred the steady state method.

Under steady state conditions, the concentration of a non-volatile tracer is given by eqn. (9.11) with $E_{OG}=0$. Providing that the tray is long enough so that tracer does not reach the liquid inlet and $Pe \gg 1$, $z \gg 0$, the solution of the equation is

$$\frac{x-x_0}{x_{in}-x_0}=\exp[-Pe(z_{in}-z)] \tag{9.20}$$

where the tracer concentration at the injection point ($z=z_{in}$) is x_{in}. Fig. 9.7 shows typical experimental tracer concentrations plotted according to eqn. (9.20) from which Pe may be determined from the slope of the lines. Using measured values of h_{cl}, eddy diffusivity is calculated from eqn. (9.21):

$$De=\frac{Q_L Z}{W h_{cl} Pe} \tag{9.21}$$

Proposed correlations for De are listed in Table 9.1 and are compared in Fig. 9.8. It can be seen that there is a wide range of predicted values of De even for air–water. Sterbacek's correlation is quite different in form from

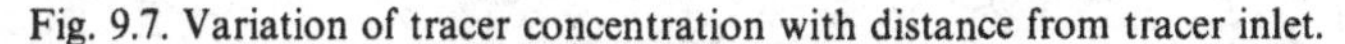

Fig. 9.7. Variation of tracer concentration with distance from tracer inlet.

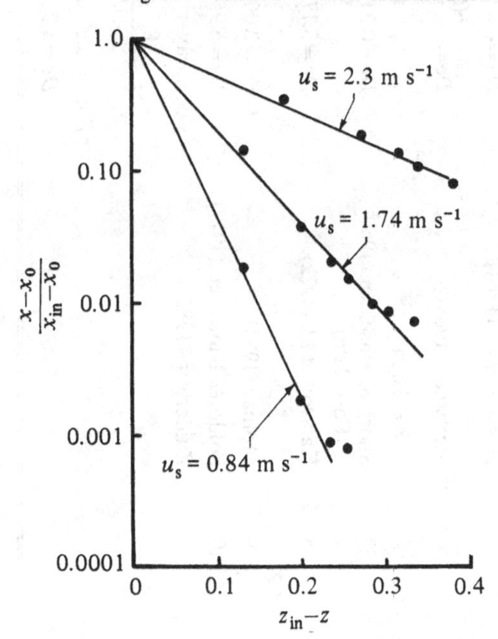

Table 9.1. *Correlations for eddy diffusivity De*

No.	Source	Correlation	Tray	Method
1	Gerster *et al.* (1958)	$De^{0.5}=0.003\,78+0.017\,u_s+3.68(Q_L/W)+0.18\,h_w$	Bubble cap	SS
2	Gilbert (1959) (Foss data)	$De=0.0025\,h_f^3h_{cl}^{-2}$	Sieve	USS
3	Barker & Self (1962)	For $h_w=0.025$ m		
		$De=3.58\times10^{-4}(Q_L/Wh_{cl})^{-0.02}h_{cl}(1-\varepsilon)^{-3.02}$	Sieve	SS
		For $h_w=0.05$–0.10 m		
		$De=1.66\times10^{-3}(Q_L/Wh_{cl})^{0.09}h_{cl}(1-\varepsilon)^{-2.91}$	Sieve	SS
4	Welch *et al.* (1964)	$De=0.088(Q_L/Wh_{cl})$	Valve	USS
5	Harada *et al.* (1964)	$De=0.0036\,h_fu_s(u_hd_h)^{-0.37}\varepsilon^{-1}$	Sieve	SS
6	Sterbacek (1968)	$De=\dfrac{0.6(Q_L/Wh_{cl})[0.57(1-\phi)^2u_h^2\rho_G+\rho_Lh_{cl}g]}{42.1[(1-\varepsilon)\rho_L+\varepsilon\rho_G]gh_{cl}}$	Sieve	USS
	$Z=0.6$ m			
7	Shore & Haselden (1969) (Foss data)	$De=0.31\,h_f[u_s(\rho_G/\rho_L)^{0.5}]^{0.63}$	Sieve	USS
8	Kafarov *et al.* (1972)	$De=3.17\times10^{-3}(Q_L/Wh_{cl})^{0.17}h_{cl}(1-\varepsilon)^{-2.83}$	Sieve with cooling coils	SS
9	Molnar (1974)	$De^{0.5}=0.0005+0.012\,85\,u_s+6.32(Q_L/W)+0.312\,h_w$	Valve	SS
10	Sohlo & Kinnunen (1977)	$De=6.95\times10^{-3}(Q_L/Wh_{cl})^{0.14}h_{cl}(1-\varepsilon)^{-2.86}$	Sieve	SS
11	Zuiderweg (1982)	For spray–mixed-froth regime		
		$De=\dfrac{8.3\,\rho_Gu_s^2h_{cl}^2}{\rho_L(Q_L/W)}$	Sieve	SS
		For emulsion regime (error in original version)		
		$De=3.0\,u_sh_{cl}(\rho_G/\rho_L)^{0.5}$	Sieve	SS

the others and involves an estimate of the tray pressure drop. Sohlo's data was obtained from a very-small-diameter tray for which the validity of eqn. (9.20) is doubtful. Gilbert's correlation is based on Foss's data (Foss 1957). Subsequently, Shore & Haselden (1969) argued that Foss misinterpreted his data and they used it to derive their own correlation. In view of these comments, it is perhaps reasonable to disregard the large predicted values of *De* shown in Fig. 9.8. Of the remaining correlations, those of Shore & Haselden and of Zuiderweg are the only ones to involve ρ_G and ρ_L and are therefore recommended for systems other than air–water. Predicted Peclét numbers are shown in Fig. 9.9 for distillation systems using these recommended correlations for a 2 m-diameter tray.

Since E_{MV} is largely independent of *Pe* when $Pe > 20$, Fig. 9.9 indicates that differences in *De* correlations are only important at low values of the flow parameter. Even then, differences are only really significant for small-diameter trays having short flow-path lengths such as was used to construct Fig. 9.9.

It has been reported that *De* for valve trays is somewhat higher than for

Fig. 9.8. Comparison of eddy diffusivity correlations. For key refer to Table 9.1. Air–water, $h_w = 0.05$ m, $d_h = 0.0127$ m, $\phi = 0.1$, $Q_L/W = 0.01$ m^3 m^{-1} s^{-1}, ε, h_{cl}, h_f taken from Figs. 3.3–3.5 using Stichlmair's correlations.

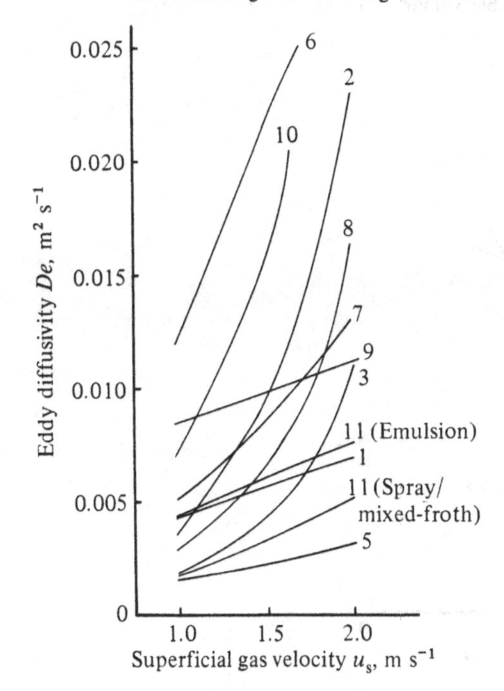

sieve trays (Biddulph 1977). This is supported by Molnar's correlation shown in Fig. 9.8.

Raper *et al.* (1984) have argued recently that *De* passes through a maximum at the froth-to-spray transition and its value falls in the spray regime. This is contrary to the prediction of all the correlations listed in Table 9.1. Raper *et al.* argued that all previous data using the steady state method were dominated by unrepresentative conditions near the exit weir, which tends to be a region of increased mixing. Further experimental work using large-diameter trays is required to test this claim.

9.8 Horizontal vapour mixing between trays

Lewis's three cases deal only with the limiting situations of either no mixing or complete mixing of vapour between trays. Partial vapour mixing has been accounted for either by a mixed vapour cells model (Ashley & Haselden 1970) or by an eddy diffusivity model (Katayama & Imoto 1972).

Providing that two or more mixed vapour cells are obtained between trays, which seems likely in industrial columns, tray efficiency is close to that calculated for unmixed vapour.

Fig. 9.9. Liquid Peclét number as a function of flow parameter. Tray dimensions, flow rates and physical properties as for Fig. 8.7. Zuiderweg's correlations (eqn. (3.26), Table 3.2) for h_{cl}, ε and hence h_f.

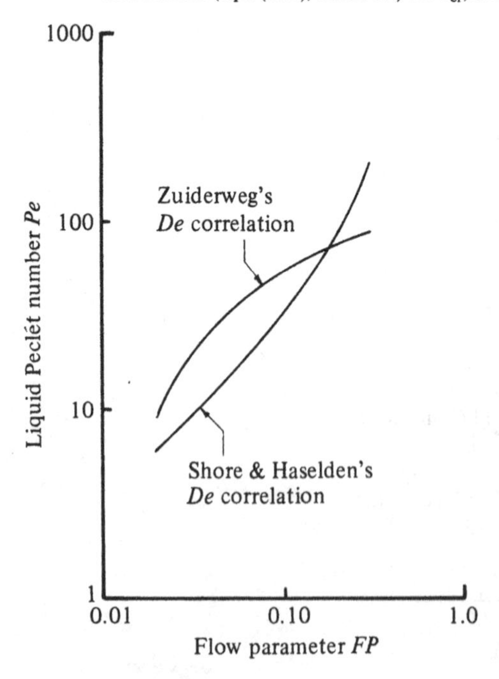

A Peclét number for horizontal vapour mixing can be defined as,

$$Pe_G = \frac{u_s Z^2}{De_G(T_s - h_f)} \tag{9.22}$$

Katayama and Imoto showed that, providing $Pe_G > 50$, the vapour can be taken as unmixed insofar as it affects tray efficiency. A typical value for De_G seems to be about 0.01 m^2 s^{-1} (Lockett *et al.* 1975) so that Pe_G is greater than 50 in all but the smallest laboratory columns. Consequently, in industrial columns, vapour can be taken as unmixed between trays. Lewis's case 1 is applicable only to the bottom tray in a column immediately above the reboiler.

9.9 Exploitation of Lewis's case 2

For typical distillation conditions ($\lambda = 1.0$, $Pe = 50$), adoption of Lewis's case 2 over case 3 gives an increase in tray efficiency of between 4 and 16 percentage points as the point efficiency varies between 60 and 80%. So a prerequisite for the use of Lewis's case 2 is that the point efficiency must be high. Some tray designs involving Lewis's case 2 are shown in Fig. 9.10. It is frequently employed in the trays used for air separation (Latimer 1967). The particular liquid flow arrangement used by Union Carbide in their parallel flow trays (Fig. 9.10) is perhaps the best-known application in the petrochemical industry. A recently claimed novel arrangement of trays giving enhanced separation, termed 'parastillation' (Jenkins 1981, Canfield 1984), is in reality no more than another example of the use of Lewis's case 2.

Fig. 9.10. Some tray designs involving Lewis's case 2.

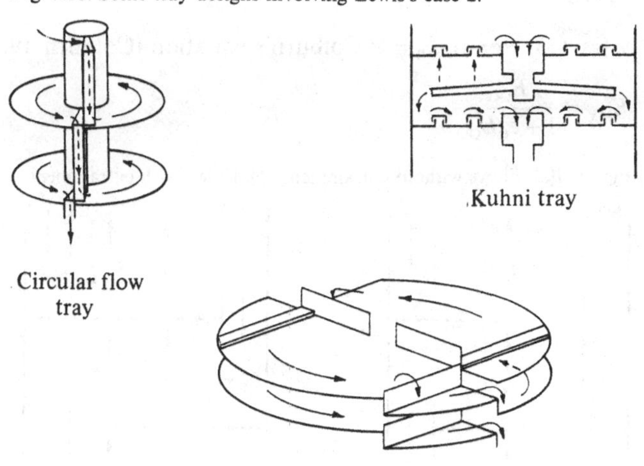

9.10 Effect of liquid entrainment on tray efficiency

Entrainment results in recycle of liquid the wrong way through the column. This causes:

- a reduction in the driving force for mass transfer between liquid and vapour,
- an increase in the liquid-to-vapour ratio on each tray,
- an increase in the liquid load and point efficiency.

The net result is usually a reduction in tray efficiency. Fig. 9.11 shows internal vapour and liquid flow rates both without and with entrainment. The apparent tray efficiency E^{a}_{MV} in the presence of entrainment is defined as (Colburn 1936, Lockett *et al.* 1983)

$$E^{a}_{MV} = \frac{\bar{Y}_n - \bar{Y}_{n-1}}{y^*_n - \bar{Y}_{n-1}} \tag{9.23}$$

where, referring to Fig. 9.12,

$$\bar{Y}_n = \bar{y}_n - \frac{E^+}{G}[\bar{x}_{n+1} - \bar{x}'_n] \tag{9.24}$$

$$\bar{Y}_{n-1} = \bar{y}_{n-1} - \frac{E^+}{G}[\bar{x}_n - \bar{x}'_{n-1}] \tag{9.25}$$

and

$$y^*_n = m\bar{x}_n + b \tag{9.26}$$

E^{a}_{MV} is used in place of the normal Murphree tray efficiency. All the effects of entrainment are rolled up into E^{a}_{MV} and it allows column simulation to be carried out independently of whether entrainment occurs or not. The following equations are available for calculation of E^{a}_{MV} with $\lambda_0 = mG/L_0$ and $e_0 = E^+/L_0$.

Lewis's cases 1–3, $Pe=0$, $\lambda_0=1$: Colburn's equation (Colburn 1936)

$$E^{a}_{MV} = \frac{E_{OG}}{1 + e_0 E_{OG}} \tag{9.27}$$

Fig. 9.11(*a*). Flows without entrainment. (*b*). Flows with entrainment.

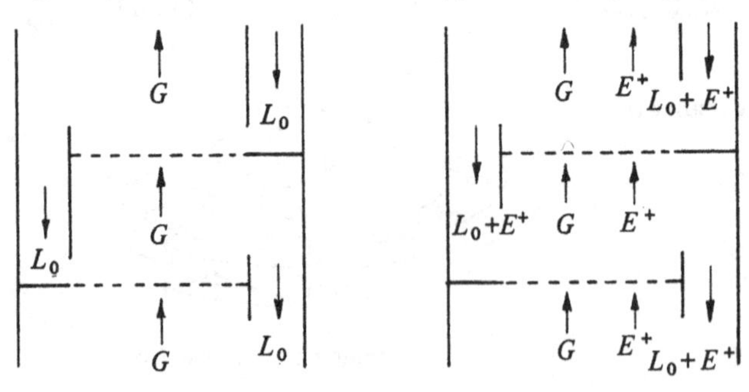

Lewis's case 2, Pe = ∞:

$$E^{a}_{MV} = \frac{\alpha - 1}{\lambda_0 - 1} \tag{9.28}$$

where α is obtained from

$$\lambda_0 = \frac{(\alpha - 1 + E_{OG})}{E_{OG}} \left[\frac{(1 + e_0) \ln \alpha}{(\alpha - 1)} - \frac{e_0}{\alpha} \right] \tag{9.29}$$

Lewis's case 3, Pe = ∞:

$$E^{a}_{MV} = \frac{\alpha - 1}{\lambda_0 - 1}$$

where if α (and λ_0) $\leqslant 1$, α is determined from

$$\lambda_0 = 2\alpha(1 + e_0) \left(\frac{A}{B^2 - C^2} \right)^{0.5} \tan^{-1} \left\{ \frac{[(B^2 - C^2)A]^{0.5}}{2\alpha(2 - E_{OG})D - (\alpha - 1 + E_{OG})(B - C)} \right\} \tag{9.30}$$

and if α (and λ_0) $\geqslant 1$, α is determined from

$$\lambda_0 = 2\alpha(1 + e_0) \left(\frac{A}{B^2 - C^2} \right)^{0.5} \tanh^{-1} \left\{ \frac{[(C^2 - B^2)A]^{0.5}}{2\alpha(2 - E_{OG})D - (\alpha - 1 + E_{OG})(B - C)} \right\} \tag{9.31}$$

where

$$A = \alpha^2 - (1 - E_{OG})^2 \tag{9.32}$$

$$B = \alpha \left[E_{OG} + \frac{e_0}{\lambda_0} (2 - E_{OG}) \right]$$

$$C = \alpha^2 \left(E_{OG} + \frac{e_0}{\lambda_0} \right) + \frac{e_0}{\lambda_0} (1 - E_{OG})$$

$$D = \alpha E_{OG} + \frac{e_0}{\lambda_0} (\alpha - 1 + E_{OG})$$

Fig. 9.12. Nomenclature for entrainment.

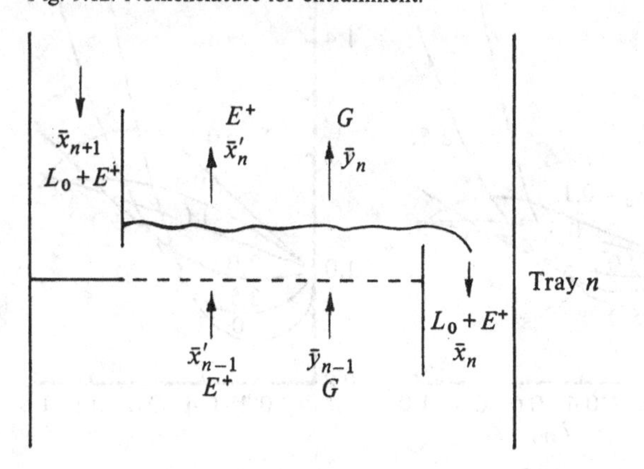

Although these equations appear fairly formidable, it turns out that α can be determined easily by simple iteration.

When $0 < Pe < \infty$, numerical solutions have been obtained using a procedure similar to that described in Section 9.5 but also allowing for entrainment using eqn. (9.7) (Lockett *et al.* 1983). Calculated values of E^{a}_{MV} for typical values of e_0, E_{OG}, Pe and λ_0 are given in Appendix B. These were calculated using both numerical solutions and eqns. (9.27)–(9.32), as appropriate.

Note that E_{OG} (and Pe) must be estimated using the increased liquid load on the tray due to recycle of liquid caused by entrainment. In principle this would appear to allow the possibility of an increase in tray efficiency with entrainment under certain circumstances.

Approximate equation for E^{a}_{MV}. A convenient and often-used approach is to replace E_{OG} by E_{MV} in eqn. (9.27) and to estimate E_{MV} taking no account of entrainment. This approach is without any theoretical foundation. Its merit is that it gives an easily calculated value for E^{a}_{MV} whose accuracy is often consistent with the accuracy to which e_0 and E_{OG} can be estimated. Fig. 9.13 shows the error involved in using this approximate method. Providing that λ_0 is close to unity, the approximate method is adequate and slightly underestimates E^{a}_{MV} compared with the rigorous methods.

9.11 Effect of weeping on tray efficiency

Calculation of the effect of weeping on tray efficiency is very similar to that described for entrainment. Weeping causes both a reduction in the

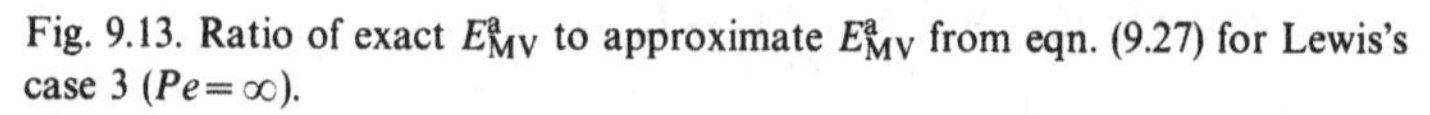

Fig. 9.13. Ratio of exact E^{a}_{MV} to approximate E^{a}_{MV} from eqn. (9.27) for Lewis's case 3 ($Pe = \infty$).

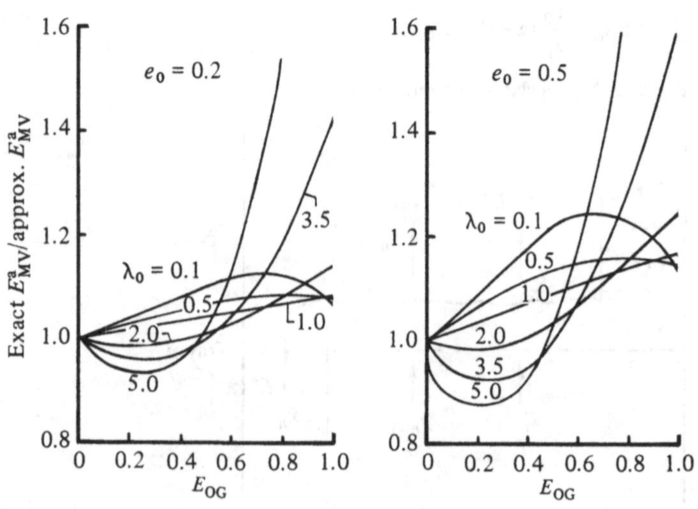

liquid weir load, Fig. 9.14, and also a reduction in tray efficiency. One difference between entrainment and weeping is that there is no convenient simple equation analogous to Colburn's equation to determine E^{a}_{MV} under weeping conditions. Using Fig. 9.15, the apparent tray efficiency E^{a}_{MV} is defined as

$$E^{a}_{MV}=\frac{\bar{Y}_n-\bar{Y}_{n-1}}{y^*_n-\bar{Y}_{n-1}} \tag{9.23}$$

where

$$\bar{Y}_n=\bar{y}_n+\frac{L_w}{G}[\bar{x}_{n+1}-\bar{x}'_{n+1}] \tag{9.33}$$

$$\bar{Y}_{n-1}=\bar{y}_{n-1}+\frac{L_w}{G}[\bar{x}_n-\bar{x}'_n] \tag{9.34}$$

and

$$y^*_n=m\bar{x}_n+b \tag{9.26}$$

Fig. 9.14. Flows with weeping.

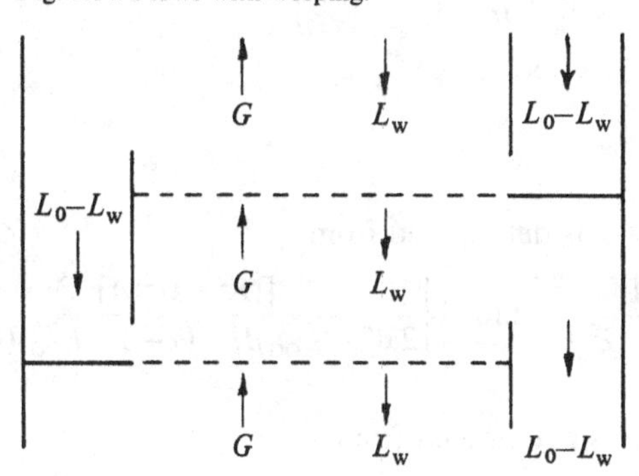

Fig. 9.15. Nomenclature for weeping.

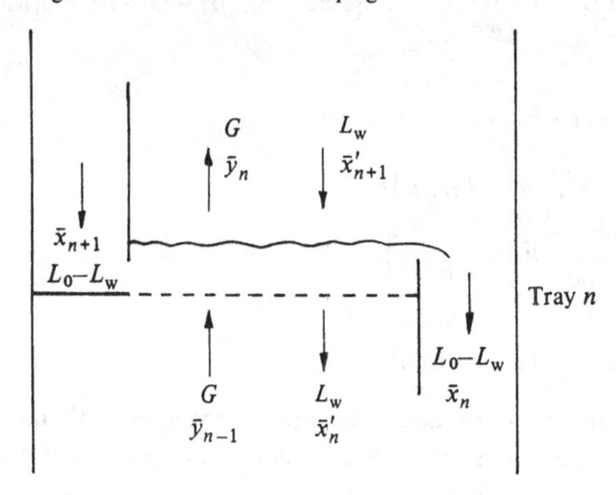

The following equations allow calculation of E^{a}_{MV} under weeping conditions. It is assumed that weeping is uniform over the tray and $\beta_0 = L_w/L_0$, $\lambda_0 = mG/L_0$.

Lewis's cases 1–3, Pe=0

$$E^{a}_{MV} = E_{OG} \tag{9.35}$$

This implies that weeping has no effect on tray efficiency providing that complete liquid mixing exists on the tray. However, because of the reduced liquid weir load and froth height caused by weeping, E_{OG} is generally reduced under weeping conditions.

Lewis's case 2, Pe=∞:

$$E^{a}_{MV} = \frac{\alpha - 1}{\lambda_0 - 1}$$

where α is obtained from

$$\lambda_0 = \left(\frac{\alpha - 1 + E_{OG}}{E_{OG}}\right)\left[\frac{(1-\beta_0)\ln\alpha}{(\alpha - 1)} + \beta_0\right] \tag{9.36}$$

Lewis's case 3, Pe=∞:

$$E^{a}_{MV} = \frac{\alpha - 1}{\lambda_0 - 1}$$

where if α (and λ_0) $\leqslant 1$, α is determined from

$$\lambda_0 = 2(1-\beta_0)\left(\frac{A}{F^2 - G^2}\right)^{0.5} \tan^{-1}\left\{\frac{[(F^2 - G^2)A]^{0.5}}{2\alpha(2 - E_{OG})H - (\alpha - 1 + E_{OG})(F - G)}\right\} \tag{9.37}$$

and if α (and λ_0) $\geqslant 1$, α is determined from

$$\lambda_0 = 2(1-\beta_0)\left(\frac{A}{G^2 - F^2}\right)^{0.5} \tanh^{-1}\left\{\frac{[(G^2 - F^2)A]^{0.5}}{2\alpha(2 - E_{OG})H - (\alpha - 1 + E_{OG})(F - G)}\right\} \tag{9.38}$$

where

$$A = \alpha^2 - (1 - E_{OG})^2 \tag{9.39}$$

$$F = E_{OG} + \frac{\beta_0}{\lambda_0}(\alpha^2 - E_{OG} + 1)$$

$$G = \alpha\left[E_{OG} + \frac{\beta_0}{\lambda_0}(2 - E_{OG})\right]$$

$$H = E_{OG} - \frac{\beta_0}{\lambda_0}(\alpha - 1 + E_{OG})$$

Calculated results from the above equations, together with numerical results for $0 < Pe < \infty$ using eqn. (9.7), are included in Appendix B (Lockett,

Rahman & Dhulesia 1984). E_{OG} should be evaluated using the reduced weir load $(L_0 - L_w)$ which results from weeping.

Non-uniform weeping. The treatment given above assumes uniform weeping over the tray. A bigger reduction in tray efficiency is obtained if weeping is localised near the inlet weir as is often the case on large trays. This has been examined by Banik (1982).

9.12 Effect of vapour entrainment on tray efficiency

Except when the liquid velocity in the downcomer is very low, some vapour is undoubtedly entrained with the liquid flowing down through the downcomer. This causes a reduction in efficiency and a simple equation for the apparent tray efficiency E^{a}_{MV} which results is (Drogaris & Lockett 1979, Lockett & Gharani 1979):

$$E^{a}_{MV} = \frac{E_{MV}}{1 + [E_{MV} m (1 - \alpha_R) \rho'_G / \alpha_R \rho'_L]} \tag{9.40}$$

E_{MV} is the Murphree tray efficiency with no vapour entrainment and α_R is the liquid volume fraction of the froth flowing under the downcomer. The assumptions leading to eqn. (9.40) are similar to those involved in the approximate version of Colburn's equation (eqn. (9.27)). Although rigorous only when $Pe = 0$, it gives an adequate estimate of E^{a}_{MV}. Application of eqn. (9.40) to published FRI data for distillation of isobutane-*n* butane, taking data from Hoek & Zuiderweg (1982), gives values as in Table 9.2.

Vapour entrainment can be significant at high pressures, and in the example above it reduces tray efficiency by up to 36%. A practical difficulty is in estimating α_R, although a good approximation is to assume $\alpha_R \approx 1.3\,\bar{\alpha}_d$ (Lockett & Gharani 1979, Zuiderweg 1982).

9.13 Effect of liquid flow maldistribution on tray efficiency

Liquid flow on a tray can be broken down into two parts. There is an underlying bulk liquid velocity profile, superimposed on which is a random movement of liquid elements caused by vapour bubbling and jetting. In combination, these determine the residence time distribution of the liquid on the tray. The simple backmixing model (Section 9.4) assumes that the underlying bulk liquid velocity profile is uniform and liquid flows across the tray as it would on a rectangular tray with no wall effects. Kirschbaum (1934) was the first to recognise that liquid flow is far from uniform and maldistribution causes a loss of efficiency.

9.13.1 *Bulk liquid velocity profiles*

As liquid flows onto a single-pass tray from the downcomer, it enters a diverging channel. It has little tendency to move sideways to follow the curved walls. Instead it tends to channel preferentially down the central part of the tray taking the shortest route from downcomer to downcomer. This leaves slower-moving, stagnant or even recirculating liquid at the sides of the tray. Examples of bulk liquid velocity profiles measured by Solari *et al.* (1982) are shown in Fig. 9.16. Solari *et al.* reported that liquid flow maldistribution became worse as the exit weir height was increased and as the gas velocity was reduced. The flow profile of Fig. 9.16*b* is identical to the one reported by Porter *et al.* (1972) and used in the stagnant regions model of Section 9.13.3. In spite of other frequent reports of liquid flow maldistribution on trays (Strand 1963, Zuiderweg *et al.* 1969, Bell 1972, Weiler *et al.* 1973, Yanagi & Scott 1973, Stichlmair & Weisshuhn 1973, Lockett & Safekourdi 1976, Neuburg & Chuang 1982), we still cannot predict the details of the liquid velocity profiles which result as tray design and hydraulic loads are changed. A contributing factor is the ease with which unsymmetrical flow patterns are formed about the axial centre line extending from weir to weir. These occur because of tray out-of-levelness, unlevel weirs and non-uniform clearance under the downcomers. Asymmetry is apparent in the liquid flow patterns reported by Bell (1972), Stichlmair & Weisshuhn (1973) and Porter *et al.* (1982). It can be especially serious on very large single-pass trays.

9.13.2 *How liquid maldistribution reduces tray efficiency*

Useful insights can be obtained without getting into the details of the mathematical models. The liquid flow profile assumed is shown in Fig. 9.17 which is similar to Fig. 9.16*b*. The liquid is taken to flow uniformly from weir to weir in the 'active' region leaving segmental 'stagnant' regions at the

Table 9.2. *Comparison of predicted and measured values of* E^{a}_{MV} ($E_{MV} = 1.4$, $m = 1$, equal molecular weights for vapour and liquid.)

Pressure (kPa)	ρ_G (kg m^{-3})	ρ_L (kg m^{-3})	α_R	E^{a}_{MV}/E_{MV} (from eqn. (9.40))	E^{a}_{MV}/E_{MV} (measured)
2070	52	437	0.36	0.77	0.76–0.79
2760	78	391	0.34	0.64	0.64–0.68

sides of the tray. This has been termed the 'stagnant regions model' (SRM) (Porter *et al.* 1972).

Since there is no bulk flow of liquid through the stagnant regions, they quickly reach equilibrium with the vapour flowing through them. Subsequent vapour passing through the stagnant regions undergoes no composition change. As the object of the tray is to change the vapour composition, this obviously reduces tray efficiency.

The explanation above is oversimplified. Although stagnant regions are not replenished by bulk liquid flow, they do receive fresh liquid from the active region by transverse liquid mixing, as depicted in Fig. 9.18. Stagnant regions tend to be depleted of lighter components due to the stripping action of the vapour and mixing from the active region replenishes them in these components. However, mixing acts only over a limited distance – estimated to be about 0.5 m by Porter *et al.* – and this distance is independent of tray diameter.

Fig. 9.16. Bulk liquid velocity profiles measured by Solari *et al.* (1982). $D = 1.25$ m, $W/D = 0.59$, $Q_L/W = 0.021\ m^3\ m^{-1}\ s^{-1}$. (*a*) $h_w = 0$, $F_s = 0.52$. (*b*) $h_w = 50$ mm, $F_s = 0.91$. (*c*) $h_w = 100$ mm, $F_s = 0.65$.

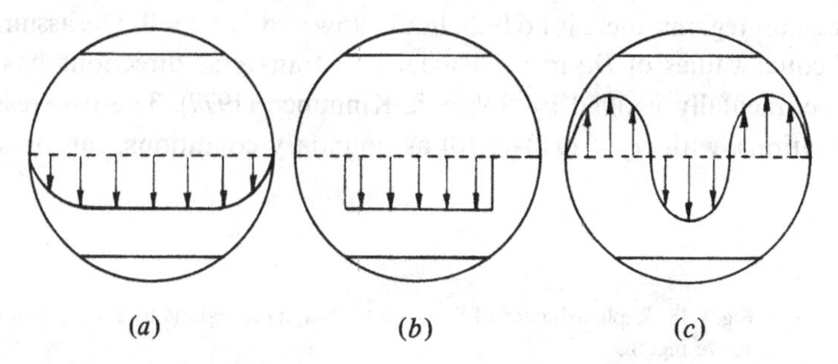

Fig. 9.17. Assumed liquid flow profile for the stagnant regions model.

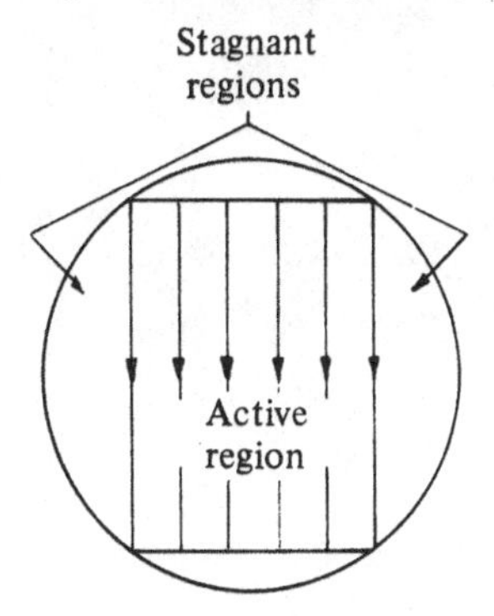

This has very important consequences for scale-up. Where the maximum width of the stagnant regions is less than about 0.5 m, transverse mixing is sufficient to overcome the adverse effects on tray efficiency. On the other hand, as tray diameter and the size of the stagnant regions increases, transverse mixing is inadequate and tray efficiency suffers as discussed above. Although this interpretation is based on the SRM, a similar argument can be used for other liquid flow patterns, e.g. those of Figs. 9.16*a*, *c* (Lockett & Safekourdi 1976).

The reduction in efficiency is even more severe when we consider a large-diameter column having single-pass trays with the stagnant regions stacked one above the other (Lockett *et al.* 1973). In principle, vapour can pass through the stagnant regions of a series of trays without undergoing significant composition change. This causes a far bigger reduction in tray and overall section efficiency than if a single tray is considered in isolation.

9.13.3 *The stagnant regions model (SRM)*

To determine the liquid concentration profile on each tray, we use eqn. (9.7) (with the simplification that $L'_w = e = 0$). In the active region, the liquid velocity profile is uniform and eqn. (9.7) holds as it stands. In the stagnant regions, there is no bulk liquid flow and $\partial x/\partial z = 0$. The assumption of equal values of *De* in both axial and transverse directions has been experimentally verified by Sohlo & Kinnunen (1977). The two resulting equations, with eqns. (9.8)–(9.10) as boundary conditions, can be solved

Fig. 9.18. Replenishment of liquid in the stagnant regions by mixing from the active region.

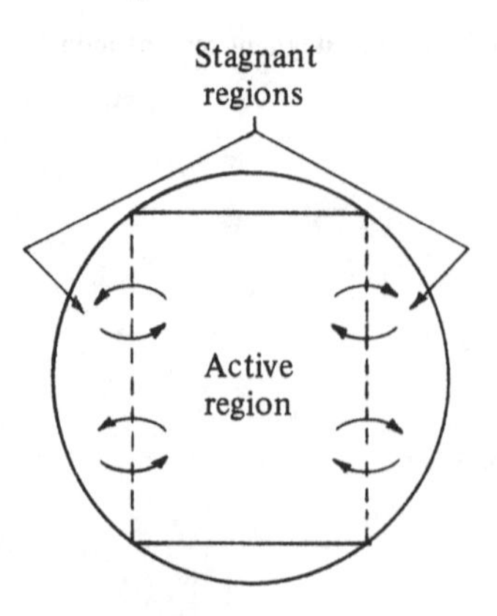

numerically (Porter *et al.* 1972) to give the liquid concentration profile on the tray. The procedure outlined in Section 9.5 is used for tray-to-tray calculation up the column. A unique feature of the numerical calculation procedure used by Porter *et al.* was the way the curved walls were taken into account while using a square mesh for the finite difference technique. For example, the wall cuts the square mesh in 17 possible ways and for each the appropriate finite difference equation had to be identified (Lim 1973).

An example of a predicted liquid concentration profile using the SRM is shown in Fig. 9.19 and it can be compared in Fig. 9.20 with an experimentally determined profile measured by Bell on an FRI tray (Bell

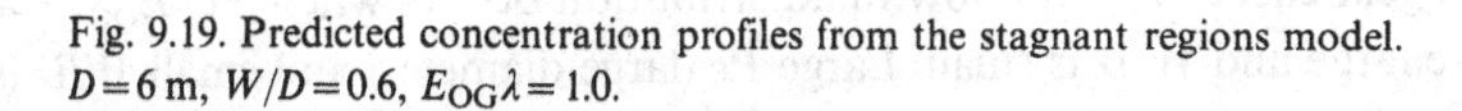

Fig. 9.19. Predicted concentration profiles from the stagnant regions model. $D = 6$ m, $W/D = 0.6$, $E_{OG}\lambda = 1.0$.

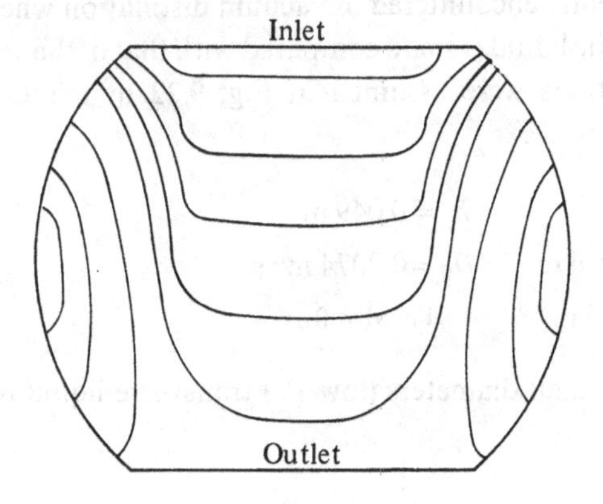

Fig. 9.20. Experimentally measured concentration profiles. $D = 2.44$ m. (Bell 1972.)

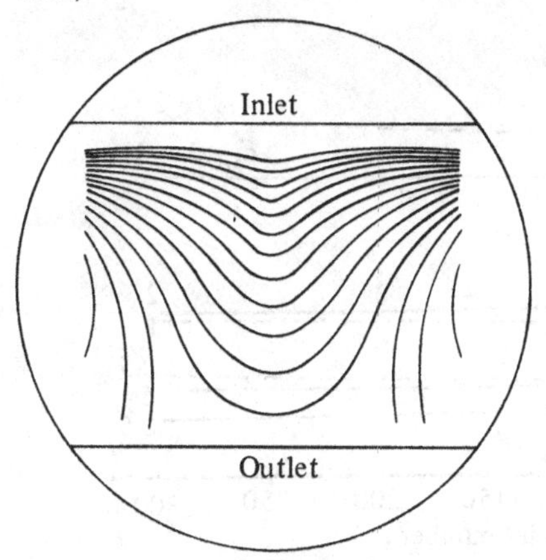

1972). Evidently, the rather sweeping assumptions regarding the liquid velocity profile made in the SRM are adequate for predicting the observed concentration profile.

The Murphree tray efficiency can be determined from the predicted liquid concentration profile using a two-dimensional version of eqn. (9.17). Tray efficiencies so obtained are compared with predictions from the simple backmixing model in Fig. 9.21 for the bottom tray in a column receiving completely mixed vapour from the reboiler (Lewis's case 1). Fig. 9.22 shows E_{MV} for a series of trays (Lewis's case 3).

Using the SRM, E_{MV} depends on Pe, E_{OG}, λ, W/D and n, where n is the tray number counting from the reboiler. Although eqn. (9.7) involves WD/A_b rather than W/D, they are easily related (Porter *et al.* 1972). The biggest effect of liquid flow maldistribution occurs when Pe, E_{OG}, λ and n are large and W/D is small. Large Pe (large diameter) and small W/D (small downcomers) are frequently encountered in vacuum distillation where the volumetric flow rate of the liquid is small compared with that of the vapour.

The following conditions were assumed in Fig. 9.22 to relate Pe to diameter:

$u_s = 0.61\ \mathrm{m\ s^{-1}}$ $\quad$ $h_{cl} = 0.049\ \mathrm{m}$

$\rho'_G = 0.044\ \text{kg-mol m}^{-3}$ $\quad$ $De = 0.0074\ \mathrm{m^2\ s^{-1}}$

$\rho'_L = 12.0\ \text{kg-mol m}^{-3}$ $\quad$ Total reflux

Fig. 9.22 shows that at small diameters (low Pe) transverse liquid mixing

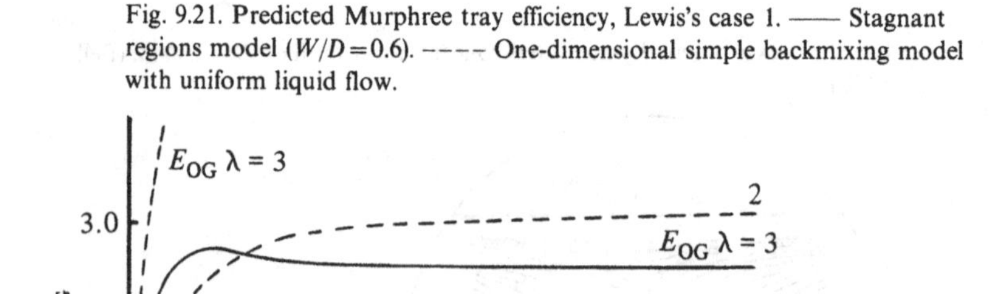

Fig. 9.21. Predicted Murphree tray efficiency, Lewis's case 1. —— Stagnant regions model ($W/D = 0.6$). ---- One-dimensional simple backmixing model with uniform liquid flow.

overcomes the adverse effects of the stagnant regions, and E_{MV} increases with diameter as it does for the simple backmixing model. Transverse liquid mixing becomes less effective at diameters above about 2 m and E_{MV} then falls with further increase in diameter. E_{MV} is predicted to pass through a maximum at a diameter of about 2 m. These numerical values depend strongly on the assumed liquid velocity profile and on the conditions used to relate *Pe* to diameter.

The SRM has been used to make predictions about two-pass trays. Liquid flow maldistribution is considerably more serious on side-to-centre flow trays than on centre-to-side flow trays (Lim *et al.* 1974, Weiler *et al.* 1981, Neuburg & Chuang 1982). Lim *et al.* predicted that the loss of efficiency would be less severe than for single-pass trays because stacking of the stagnant regions, one above the other, is avoided. Neuburg & Chuang measured liquid flow maldistribution on two-pass trays and used the results to predict, with some success, the efficiency of trays used in the Girdler-Sulfide process for heavy water production.

9.13.4 *Other liquid maldistribution models*

Although the SRM is oversimplified in its representation of the bulk liquid velocity profile, its major advantage is that it allows ready understanding of the principles involved, particularly in relation to scale-

Fig. 9.22. Predicted Murphree tray efficiencies using the stagnant regions model – Lewis's case 3 for $n > 1$.

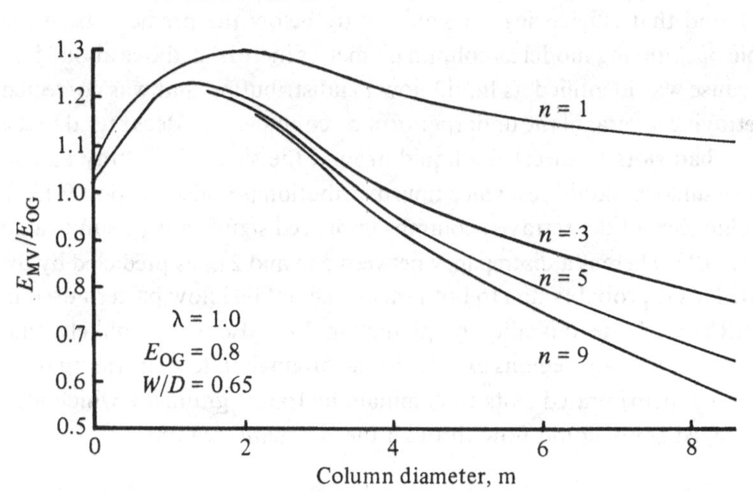

up. The results are expressed in dimensionless groups whose values are easily calculated from existing correlations. In principle any bulk liquid velocity profile can be used in conjunction with eqn. (9.7) at the expense of additional complexity – see Lockett & Safekourdi 1976, Weiler *et al.* 1981, Lim *et al.* 1974, for examples. More complicated models have divided the tray into zones of forward flow and recirculating flow with a certain fraction of the liquid being recirculated. Unfortunately, this introduces additional parameters which are difficult to predict (Strand 1963, Bell & Solari 1974, Brambilla 1976, Kafarov *et al.* 1979, Kuotsung *et al.* 1983). The models have been refined to include such things as unequal values of eddy diffusivity in the axial and transverse directions (Lockett & Safekourdi 1976, Solari & Bell 1978, Sohlo & Kouri 1982), mixing in the downcomer (Weiler *et al.* 1981, Lockett & Banik 1981) and development of the liquid velocity profile (Kouri & Sohlo 1985). These refinements are of questionable worth until the underlying bulk liquid velocity profiles can be predicted. In general the predictions of all these models differ only in minor detail from those of the SRM.

9.13.5 *Experimental evidence for efficiency reduction due to liquid flow maldistribution*

The results shown in Fig. 9.23 relate to Union Carbide slotted sieve trays in vacuum distillation service, and they vividly demonstrate the effect that liquid flow maldistribution has on the efficiency of large-diameter columns. Early designs of these trays took no account of liquid flow pattern and used slots in the tray deck simply to overcome hydraulic gradient. It was found that efficiencies fell significantly below the predictions of the simple backmixing model as column diameters increased above about 5 m. The cause was identified as liquid flow maldistribution and was corrected by retraying several of the underperforming columns. The decks used in the retrays had slots to direct the liquid around the sides of the tray and so achieve uniform liquid residence time distribution (see also Section 9.13.6). The efficiency of the retrayed columns improved significantly as indicated by Fig. 9.23. There is a discrepancy between 5 m and 2 m as predicted by the SRM. This is probably due to both the oversimplified flow pattern used in the SRM and the difficulty of predicting *Pe*. Also it is unlikely that completely stagnant regions existed on the original underperforming trays since they incorporated slots to eliminate hydraulic gradients which also encouraged some liquid flow through the 'stagnant' regions.

FRI (Yanagi & Scott 1973) has reported some distillation results which purported to examine the influence of liquid maldistribution on efficiency by testing small-diameter trays with and without liquid flow control. The conditions of the tests were such that liquid flow maldistribution should not have had a noticeable effect on efficiency according to the predictions of the SRM. No effect on efficiency was indeed found – a somewhat unimpressive confirmation of the theory.

9.13.6 *Control of the liquid velocity profile*

In medium- to large-diameter columns, where the effects of liquid flow maldistribution on efficiency can be severe, liquid flow control is worthwhile. It can be accomplished by slots in the tray deck, arranged so that the momentum imparted to the liquid via the vapour passing through them is used to direct the liquid and so achieve uniform liquid residence time distribution (Smith & Delnicki 1975). Baffles, cut-down weirs near the walls (Yanagi & Scott 1973), and curved downcomers have also been suggested to improve the liquid flow pattern.

Providing that uniform liquid residence time distribution is achieved, the simple backmixing model can be used to predict E_{MV}/E_{OG}. To allow for the curved walls, even though the model is derived for a rectangular tray, the

Fig. 9.23. Ratio of observed to calculated tray efficiency vs. column diameter for large single-pass vacuum distillation trays. (Smith & Delnicki 1975.)

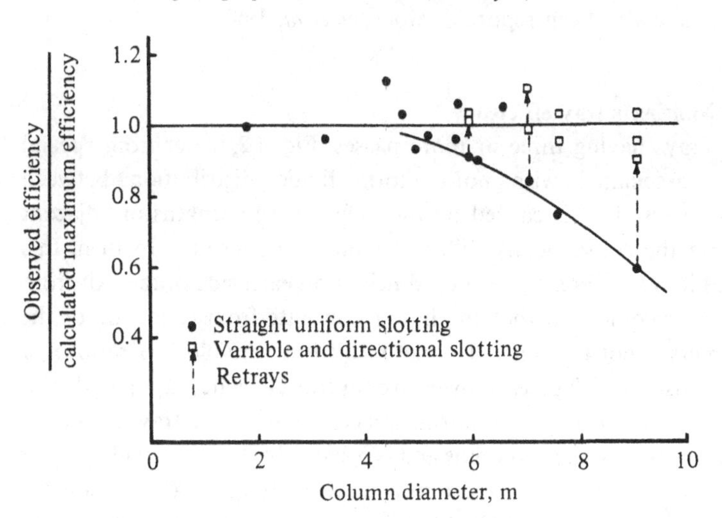

Peclét number should be based on the effective width, S, of the circular tray, which for single-pass trays is given by

$$\frac{S}{D}=0.5\frac{W}{D}+\frac{(0.5\pi/180)\sin^{-1}\{[1-(W/D)^2]^{0.5}\}}{[1-(W/D)^2]^{0.5}} \tag{9.41}$$

and

$$Pe=\frac{Q_L Z}{Sh_{cl} De} \tag{9.42}$$

The calculated efficiency of a rectangular tray of width S and flow path length Z, using the simple backmixing model, corresponds to the efficiency of a circular single-pass tray having a uniform liquid residence time distribution, for which the flow path length is also Z, the weir length is W and the diameter is D (Lockett & Banik 1981).

9.13.7 *Non-uniform vapour flow*

Because of hydraulic gradient, there is a tendency for excess vapour to flow through the tray near the exit weir. It can cause local entrainment and reduced point efficiency because of the reduction in vapour contact time. This type of vapour maldistribution, however, has only a very minor effect on the ratio E_{MV}/E_{OG} (Lockett & Dhulesia 1980). Vapour maldistribution perpendicular to the direction of liquid flow causes a much greater reduction in tray efficiency (Vybornov *et al.* 1971). This can result from tray out-of-levelness or column tilt. Studies of the effect on tray efficiency of tray oscillation and tilt, such as are found in offshore-based columns, have also been reported (Hoerner *et al.* 1982).

9.14 Multipass tray efficiency

Trays having three or more passes, Fig. 1.2, suffer from special problems associated with non-uniform liquid distribution between different passes. This is caused by the different weir lengths of adjacent passes and the consequently different liquid crest heights. In turn, this translates into a different clear liquid height on each side of the exit from interior downcomers so that the liquid flow rate from each side of the downcomers is not the same. The net result is a variable L/G ratio over different passes, which gives a lower overall tray efficiency compared with the situation where L/G is uniform everywhere over the tray. A similar situation exists in packed columns and is responsible for the usually poorer separation performance of randomly packed columns as the column diameter increases. Bolles (1976*b*) has described a model which deals with the hydrodynamics and efficiency of multipass trays and has given a number of recommendations for achieving uniform L/G. In summary, these are:

- Remove control of liquid flow from the exit weirs by using inlet weirs which, of necessity, are of equal length on each side of interior downcomers; this ensures equal L in each pass.
- Use the same open hole area in each pass to give equal values of G (providing that the dry tray pressure drop dominates the overall tray pressure drop).

In the absence of these special measures, multipass tray efficiency can be up to 40% lower than normal.

10

Prediction of efficiency for multicomponent mixtures

10.1 Introduction

Although the two component efficiencies in a binary mixture are identical, in a multicomponent mixture the component efficiencies are usually all different. This arises because, in both the vapour and liquid phases, each component has a different diffusion coefficient through the mixture. Diffusional interactions also play a part. Furthermore, the effective slope of the equilibrium line m differs for each component, so each has a different percentage liquid-phase resistance and a different ratio of E_{MV}/E_{OG}. There is a fairly large and growing body of experimental work in which different component efficiencies have been measured and this has been summarised by Krishna and others (Krishna *et al.* 1977, Krishna & Standart 1979, Vogelpohl 1979, Medina *et al.* 1979*a*, *b*, Aittamaa 1981, Chan & Fair 1984).

The differences in component point efficiencies tend to be small when the mixtures are ideal and the species are similar in size and nature. Even then, because of different m values, tray efficiencies can differ (Biddulph 1975*b*). For non-ideal mixtures both point and tray efficiencies can vary widely between components. The presence of water in an organic mixture is a common situation where different component efficiencies can be expected.

Toor & Burchard (1960) gave an example for the distillation of a mixture of methanol–isopropanol–water in which it was calculated that 84 trays were required to achieve a given separation assuming equal component tray efficiencies of 40%. Taking into account differences in diffusion coefficients and diffusional interactions resulted in 117 trays being needed to achieve the same separation.

The possibility of undertraying a column by such a large extent is usually sufficient to capture the interest of design engineers in this topic. One reason that individual component efficiencies are not routinely predicted at the design stage is that the prediction methods have not been sufficiently tested against data from industrial columns. Another is that theoretical treatments of the subject often make formidable reading for the average process engineer and it is not easy to pick out a design procedure from the information provided. Here the intention is to record three design methods which have been proposed for efficiency prediction in multicomponent mixtures. The aim is to aid in the transfer of more soundly based prediction methods from the university laboratory to the design office and eventually to the plant.

10.2 Pseudo-binary approach – method 1

The treatment of a multicomponent mixture as a pseudo-binary mixture based on two key components is by far the most common procedure used in practice. Although not rigorous, its attraction is that the relatively simple methods developed for binary mixtures can be retained with a minimum of change. Some judgement is required in selecting the keys and the two components selected are often not the same in different parts of the column. For example, for a feed consisting of components A, B and C, listed in order of decreasing volatility, A and B might be the keys above the feed and B and C below. The choice depends on the feed and product compositions involved and the volatility differences between the components. The steps in the procedure are:

(i) From a column simulation in terms of theoretical trays, locate representative trays in each section of the column.

(ii) For the mixture as a whole, predict $\mu_G, \mu_L, \rho_G, \rho_L$ and if necessary σ. Recommended methods have been documented by Reid *et al.* (1977).

(iii) Select light and heavy key components and determine the composition of the pseudo-binary mixture as

$$Y_{lk} = \frac{y_{lk}}{y_{lk} + y_{hk}}, \quad X_{lk} = \frac{x_{lk}}{x_{lk} + x_{hk}} \tag{10.1}$$

(iv) Predict the binary diffusion coefficients for the keys in each phase at the temperature and pressure of the mixture (Reid *et al.* 1977).

(v) Calculate Y_{lk} and X_{lk} on adjacent theoretical trays $n+1$ and n, and determine m from

$$m = \frac{(Y_{lk})_{n+1} - (Y_{lk})_n}{(X_{lk})_{n+1} - (X_{lk})_n} \tag{10.2}$$

(vi) Use the correlations and equations developed for binary mixtures to predict E_{OG}, E_{MV} and E_0.

The section efficiency E_0 is used to determine the number of actual trays in each section of the column, when used in conjunction with a theoretical tray column simulation program. However, because of differences in component efficiencies, terminal and side draw product compositions will not necessarily be the same as given by a theoretical tray simulation. An additional step which is sometimes used is to assume that each component has the same value of E_{MV} as calculated for the pseudo-binary mixture, and to resimulate the column as discussed after (vi) of method 2 below.

10.3 Prediction of individual component efficiencies with no diffusional interactions – method 2

This procedure, outlined in the *Bubble Tray Design Manual* (AIChE 1958) yields individual component efficiencies but takes no account of diffusional interactions. The steps are:

(i) Predict mixture physical properties as in (i) and (ii) of Section 10.2.

(ii) Predict the gas-phase diffusion coefficient $(D_G)_{ij}$ for all possible binary pairs in the mixture (Reid *et al.* 1977).

(iii) Predict the gas-phase diffusion coefficient for each component i in the presence of all the other components, e.g. using the Wilke equation (Wilke 1950):

$$(D_G)_i = \frac{1 - y_i}{\sum_{j=1 \neq i}^{n} [y_j/(D_G)_{ij}]} \tag{10.3}$$

(iv) In the liquid phase, predict the diffusion coefficient, $(D_L)_i$, of each component as solute through a mixture of the other components as solvent (Reid *et al.* 1977).

(v) Obtain y and x for trays $n+1$ and n from a theoretical tray simulation and determine m_i for each component from

$$m_i = \frac{(y_i)_{n+1} - (y_i)_n}{(x_i)_{n+1} - (x_i)_n} \tag{10.4}$$

(vi) Calculate E_{OGi} and E_{MVi} for each component using the equations developed for binary mixtures.

The procedure yields a different tray efficiency E_{MVi} for each component but ignores diffusional interactions. Overall section efficiencies have no meaning since to calculate them would imply a different number of actual trays for each component. Instead the calculated E_{MVi} should be used in a column simulation program which has provision to accept individual component efficiencies on each tray. A simplified version of this procedure was adopted by Biddulph (1975*b*) for the simulation of an air separation column. Based on experimental evidence, he assumed identical E_{OGi} for oxygen, nitrogen and argon but calculated different E_{MVi} because of the different m_i values for the three components.

10.4 Prediction of individual component efficiencies including diffusional interactions –method 3

This method also leads to the calculation of individual component efficiencies but takes into account diffusional interactions. The calculation procedure is based on the Maxwell–Stefan equations for diffusion. The derivation of the equations is given in articles by Diener & Gerster (1968), Krishna *et al.* (1977) and Krishna (1977). The following assumptions are made:

- No influence of finite mass-transfer rates on the mass-transfer coefficients.
- Thermodynamic correction factors may be neglected.
- Equimolar mass transfer.

Krishna (1977) has indicated how these assumptions can be relaxed, but in so doing the complexity of the calculations increases considerably. The equations are given for a ternary mixture. More general matrix equations can be found in the original references. The calculation procedure is as follows:

(i) Estimate the three binary diffusion coefficients (D_{12}, D_{13}, D_{23}) in each phase using, for example, the methods given by Reid *et al.* (1977).

(ii) Using one of the correlations developed for binary mixtures, determine the number of transfer units for each of the three binaries in each phase, $(N_G)_{ij}$ and $(N'_L)_{ij}$ ($ij = 12, 13, 23$). (Note: The theory has not yet been developed for $(N_L)_{ij}$ replacing $(N'_L)_{ij}$.)

(iii) Calculate the number of ternary transfer units in the gas phase $(\mathcal{N}_G)_{ij}$ ($ij = 11, 12, 21, 22$) from:

$$\begin{aligned}
(\mathcal{N}_G)_{11} &= (N_G)_{13}[y_1(N_G)_{23} + (1-y_1)(N_G)_{12}]/\psi \\
(\mathcal{N}_G)_{12} &= y_1(N_G)_{23}[(N_G)_{13} - (N_G)_{12}]/\psi \\
(\mathcal{N}_G)_{21} &= y_2(N_G)_{13}[(N_G)_{23} - (N_G)_{12}]/\psi \\
(\mathcal{N}_G)_{22} &= (N_G)_{23}[y_2(N_G)_{13} + (1-y_2)(N_G)_{12}]/\psi
\end{aligned} \qquad (10.5)$$

where

$$\psi = y_1(N_G)_{23} + y_2(N_G)_{13} + y_3(N_G)_{12}$$

(iv) Calculate the number of ternary transfer units in the liquid phase $(\mathcal{N}'_L)_{ij}$ ($ij = 11, 12, 21, 22$) from eqn. (10.5) with $(\mathcal{N}_G)_{ij}$ replaced by $(\mathcal{N}'_L)_{ij}$, $(N_G)_{ij}$ replaced by $(N'_L)_{ij}$ and y_i replaced by x_i.

(v) Determine m values from a theoretical tray column simulation using compositions taken from two adjacent trays $n+1$ and n:

$$\begin{aligned} m_{11} &= \frac{(y_1)_{n+1} - (y_1)_n}{(x_1)_{n+1} - (x_1)_n} \\ m_{12} &= \frac{(y_1)_{n+1} - (y_1)_n}{(x_2)_{n+1} - (x_2)_n} \\ m_{21} &= \frac{(y_2)_{n+1} - (y_2)_n}{(x_1)_{n+1} - (x_1)_n} \\ m_{22} &= \frac{(y_2)_{n+1} - (y_2)_n}{(x_2)_{n+1} - (x_2)_n} \end{aligned} \tag{10.6}$$

(vi) Combine the number of ternary transfer units in each phase to obtain the number of ternary overall gas-phase transfer units $(\mathcal{N}_{OG})_{ij}$ ($ij=$ 11, 12, 21, 22) using eqn. (10.7)

$$\begin{aligned} \frac{(\mathcal{N}_{OG})_{11}}{\alpha_{OG}} &= \frac{(\mathcal{N}_G)_{11}}{\alpha_G} + \left[\frac{m_{22}(\mathcal{N}'_L)_{11} - m_{21}(\mathcal{N}'_L)_{12}}{\alpha_L}\right]\left(\frac{G}{L}\right) \\ \frac{(\mathcal{N}_{OG})_{12}}{\alpha_{OG}} &= \frac{(\mathcal{N}_G)_{12}}{\alpha_G} + \left[\frac{m_{11}(\mathcal{N}'_L)_{12} - m_{12}(\mathcal{N}'_L)_{11}}{\alpha_L}\right]\left(\frac{G}{L}\right) \\ \frac{(\mathcal{N}_{OG})_{21}}{\alpha_{OG}} &= \frac{(\mathcal{N}_G)_{21}}{\alpha_G} + \left[\frac{m_{22}(\mathcal{N}'_L)_{21} - m_{21}(\mathcal{N}'_L)_{22}}{\alpha_L}\right]\left(\frac{G}{L}\right) \\ \frac{(\mathcal{N}_{OG})_{22}}{\alpha_{OG}} &= \frac{(\mathcal{N}_G)_{22}}{\alpha_G} + \left[\frac{m_{11}(\mathcal{N}'_L)_{22} - m_{12}(\mathcal{N}'_L)_{21}}{\alpha_L}\right]\left(\frac{G}{L}\right) \end{aligned} \tag{10.7}$$

where

$$\begin{aligned} \alpha_{OG} &= (\mathcal{N}_{OG})_{11}(\mathcal{N}_{OG})_{22} - (\mathcal{N}_{OG})_{12}(\mathcal{N}_{OG})_{21} \\ \alpha_G &= (\mathcal{N}_G)_{11}(\mathcal{N}_G)_{22} - (\mathcal{N}_G)_{12}(\mathcal{N}_G)_{21} \\ \alpha_L &= (\mathcal{N}'_L)_{11}(\mathcal{N}'_L)_{22} - (\mathcal{N}'_L)_{12}(\mathcal{N}'_L)_{21} \end{aligned}$$

(vii) The elements G_{ij} ($ij = 11$, 12, 21, 22) are next evaluated from:

$$\begin{aligned} G_{11} &= \exp(a)\left\{\left[\frac{(\mathcal{N}_{OG})_{22} - (\mathcal{N}_{OG})_{11}}{2b}\right]\sinh b + \cosh b\right\} \\ G_{12} &= -\exp(a)\left[\frac{(\mathcal{N}_{OG})_{12}\sinh b}{b}\right] \\ G_{21} &= -\exp(a)\left[\frac{(\mathcal{N}_{OG})_{21}\sinh b}{b}\right] \\ G_{22} &= \exp(a)\left\{\left[\frac{(\mathcal{N}_{OG})_{11} - (\mathcal{N}_{OG})_{22}}{2b}\right]\sinh b + \cosh b\right\} \end{aligned} \tag{10.8}$$

where

$$a=-\left[\frac{(\mathcal{N}_{OG})_{11}+(\mathcal{N}_{OG})_{22}}{2}\right]$$

$$b=\left\{\frac{[(\mathcal{N}_{OG})_{11}-(\mathcal{N}_{OG})_{22}]^2}{4}+(\mathcal{N}_{OG})_{21}(\mathcal{N}_{OG})_{12}\right\}^{0.5}$$

(viii) The point efficiencies in the ternary mixture are defined as

$$(E_{OGi})_n=\frac{y_{in}-y_{in-1}}{y^*_{in}-y_{in-1}} \tag{10.9}$$

where y^*_{in} is the mole fraction of i in the vapour which is in equilibrium with the liquid at a point on the tray. Point efficiencies may be calculated from:

$$E_{OG1}=1-G_{11}-\frac{G_{12}}{\gamma}$$
$$E_{OG2}=1-G_{22}-G_{21}\gamma \tag{10.10}$$
$$E_{OG3}=\frac{\gamma E_{OG1}+E_{OG2}}{\gamma+1}$$

where

$$\gamma=\frac{y^*_{1n}-y_{1n-1}}{y^*_{2n}-y_{2n-1}}$$

Initially, γ can conveniently be approximated by

$$\gamma=\frac{m_{11}}{m_{21}} \quad \text{or} \quad \gamma=\frac{m_{12}}{m_{22}} \tag{10.11}$$

and a better estimate can subsequently be made once the concentration profiles through the column have been established.

(ix) Murphree tray efficiencies E_{MV1}, E_{MV2}, E_{MV3} can be determined from the point efficiencies using one of the mixing models outlined in Chapter 9. The slope of the equilibrium line used should be m_{11} for E_{MV1} and m_{22} for E_{MV2} from eqn. (10.6) with a corresponding expression for m_{33} used for E_{MV3}.

Krishna has argued that the use of eqn. (10.5) to predict $(\mathcal{N}'_L)_{ij}$ from binary data as in (iv) is inaccurate because it neglects thermodynamic interactions. He has developed procedures to overcome this limitation (Krishna 1976, Krishna & Standart 1979). Even then, calculation of $(\mathcal{N}'_L)_{ij}$ is not straightforward because of the difficulty of predicting appropriate liquid-phase diffusion coefficients in a multicomponent mixture. Perhaps because of these difficulties, when these equations have been tested against experimental data, it has often been assumed that mass transfer resistance resides entirely in the vapour phase. Using this assumption, Medina *et al.* (1979*b*) and Vogelpohl (1979) were able to predict ternary distillation column composition profiles using binary data. Similarly, Krishna *et al.* (1977) claimed that the trends shown by their measured ternary point efficiencies were in agreement with the predictions of the model with liquid-phase resistance neglected. On the other hand, Aittamaa (1981) obtained good agreement with measured column composition profiles using the full theory given above including liquid-phase resistances. In view of the discussion given in Section 8.5.3, it is doubtful whether liquid-phase resistance can be neglected in general.

Because of their dependency on γ (eqn. (10.10)), point efficiencies can take on values outside the range 0–1.0, and even negative values can be obtained. This is particularly possible for components of intermediate volatility in the region of the column where their composition goes through a maximum or minimum. The question of how one uses the mixing models to calculate E_{MV} from negative values of E_{OG} has not been resolved. Fortunately, Medina *et al.* (1979*b*) have pointed out that predicted composition profiles are not sensitive to the precise values of the efficiencies in regions of concentration maxima or minima.

10.5 Numerical example

The following section demonstrates the application of each of the three prediction methods to a real case involving the distillation of methanol–ethanol–water. At the outset we must decide on which of the prediction methods for binary efficiencies will be adapted to this multicomponent situation. As shown earlier, all binary prediction methods except that of Zuiderweg underpredict the liquid-phase resistance to mass transfer. If we were to use them in the present case, the differences in the procedures for dealing with multicomponent mass transfer in the liquid phase would have no significant impact on the predicted efficiencies, so there seems to be a case for using Zuiderweg's method here. However, a problem arises because Zuiderweg's prediction method for N_G does not involve the gas-phase diffusion coefficient and so predicts no difference in N_G for the three components. The approach we shall adopt is to use the AIChE method for N_G and Zuiderweg's method for N'_L. As shown in Fig. 8.7, the AIChE method and Zuiderweg's method give similar predictions for N_G even though the former involves gas-phase diffusivities whereas the latter does not. The following are obtained from a column simulation in terms of theoretical trays:

Compositions and general data

	Methanol	Ethanol	Water
y in vapour leaving tray n	0.913 59	0.17780×10^{-2}	0.084 63
y in vapour leaving tray $n-1$	0.890 74	$0.162\,39 \times 10^{-2}$	0.107 64
x in liquid leaving tray n	0.843 19	$0.217\,99 \times 10^{-2}$	0.154 63
x in liquid leaving tray $n-1$	0.798 96	$0.189\,19 \times 10^{-2}$	0.199 15
Component μ_L cP	0.17	0.28	0.24
Molecular weight M	32.04	46.07	18.02
Association factor ϕ_A	1.9	1.5	2.6
Liquid molar volume at normal boiling point V_i cm^3/g mol	42.1	62.5	18.9

For tray n. Pressure = 4.70 bar, temperature $T_L = 385.4$ K, $\mu_G = 0.984 \times 10^{-5}$ N s m^{-2}, μ_L mixture = 0.18 cP, $\sigma = 18.3 \times 10^{-3}$ N m^{-1}, $\rho_G = 4.93$ kg m^{-3}, $\rho_L = 717$ kg m^{-3}.

Hydraulic conditions. $F_s = 1.93$ m s^{-1}(kg m^{-3})$^{0.5}$, $Q_L/W = 1.28 \times 10^{-2}$ m^3 m^{-1} s^{-1}, $u_s = 0.87$ m s^{-1}, $h_w = 0.05$ m, $\phi = 0.1$, $h_{cl} = 0.035$ m, total reflux is assumed hereafter to simplify calculation.

Gas-phase binary diffusion coefficients are calculated by Brokaw's method since these are polar components (Reid *et al.* 1977): $(D_G)_{ME} = 2.53 \times 10^{-6}$, $(D_G)_{MW} = 4.75 \times 10^{-6}$, $(D_G)_{EW} = 3.94 \times 10^{-6}$(m^2 s^{-1}).

Liquid-phase binary diffusion coefficients at infinite dilution are calculated by the Wilke–Chang equation (Reid *et al.* 1977) which involves pure component liquid viscosities:

$$(D_L^0)_{ME} = 8.98 \times 10^{-9}, \quad (D_L^0)_{EM} = 1.09 \times 10^{-8}$$

$$(D_L^0)_{MW} = 8.62 \times 10^{-9}, \quad (D_L^0)_{WM} = 2.25 \times 10^{-8}$$

$$(D_L^0)_{EW} = 6.80 \times 10^{-9}, \quad (D_L^0)_{WE} = 1.45 \times 10^{-8} \quad (\text{m}^2\,\text{s}^{-1})$$

Method 1 – pseudo-binary approach

(iii) Methanol and ethanol are the respective light and heavy keys. From eqn. (10.1), $(Y_M)_n = 0.99806$, $(X_M)_n = 0.99742$, $(Y_M)_{n-1} = 0.99818$, $(X_M)_{n-1} = 0.99764$.

(iv) As an approximation, we use the Vignes equation (Reid *et al.* 1977) to correct the liquid-phase diffusivities for concentration: $(D_L)_{ME} = (D_L^0)_{ME}^{X_E} \cdot (D_L^0)_{EM}^{X_M} = 1.09 \times 10^{-8}$ m^2 s^{-1}; also $(D_G)_{ME} = 2.53 \times 10^{-6}$ m^2 s^{-1}.

(v) From eqn. (10.2), $m = 0.545$.

(vi) Using eqn. (8.48), $N_G = 2.13$. Using eqns. (8.45) and (8.47) and, since at total reflux, $N'_L = k'_L a h_f \rho_L/(u_s \rho_G)$, $N'_L = 2.67$. At total reflux $\lambda = 0.545$ and from eqns. (8.23) and (8.29), $E_{OG} = 77\%$. As a simplification, we will use the simplest liquid mixing model, eqn. (9.1), assuming Lewis's case 1 and large *Pe*, so that $E_{MV} = 96\%$.

Method 2 – no diffusional interactions

(iii) Using eqn. (10.3) the vapour-phase diffusivities ($m^2 s^{-1}$) are $(D_G)_M = 4.67 \times 10^{-6}$, $(D_G)_E = 2.61 \times 10^{-6}$, $(D_G)_W = 4.75 \times 10^{-6}$.

(iv) The liquid diffusion coefficient of component *i* diffusing through the liquid mixture is given approximately by a modified form of the Wilke–Chang equation:

$$(D_L^0)_i = \frac{7.4 \times 10^{-12} (\phi_A M)_i^{0.5} T_L}{\mu_L V_i^{0.6}} \quad (m^2 s^{-1}) \tag{10.12}$$

For convenience μ_L is taken as the mixture viscosity (cP) and $\phi_A M$ is estimated using the suggestion of Perkins and Geankoplis (Reid *et al.* 1977):

$$(\phi_A M)_i = \sum_{\substack{j=1 \\ j \neq i}}^{n} (x_j \phi_{Aj} M_j)$$

From these equations, $(D_L^0)_M = 4.57 \times 10^{-9}$, $(D_L^0)_E = 1.01 \times 10^{-8}$, $(D_L^0)_W = 1.95 \times 10^{-8}$ ($m^2 s^{-1}$). There is no clear-cut recommended way of correcting these $(D_L^0)_i$ values for concentration in this method and so it is ignored.

(v) From eqn. (10.4), $m_M = 0.517$, $m_E = 0.535$, $m_W = 0.517$.

(vi) Using the equations itemised under (vi), Method 1, yields:

for methanol, $N_G = 2.88$, $N'_L = 2.14$, $E_{OG} = 82\%$, $E_{MV} = 102\%$

for ethanol, $N_G = 2.16$, $N'_L = 2.62$, $E_{OG} = 78\%$, $E_{MV} = 96\%$

for water, $N_G = 2.91$, $N'_L = 3.09$, $E_{OG} = 86\%$, $E_{MV} = 108\%$

Method 3 – allowing for diffusional interactions

(i) We use a suggested modification of the Vignes equation due to Krishna to correct the liquid-phase binary diffusion coefficients at infinite dilution for concentration:

$$(D_L)_{ij} = (D_L^0)_{ij}^{x_j/(x_i+x_j)} \cdot (D_L^0)_{ji}^{x_i/(x_i+x_j)} \tag{10.14}$$

Using eqn. (10.14) gives $(D_L)_{ME} = 1.09 \times 10^{-8}$, $(D_L)_{MW} = 1.94 \times 10^{-8}$, $(D_L)_{EW} = 6.87 \times 10^{-9}$ ($m^2 s^{-1}$). In addition, $(D_G)_{ME} = 2.53 \times 10^{-6}$, $(D_G)_{MW} = 4.75 \times 10^{-6}$, $(D_G)_{EW} = 3.94 \times 10^{-6}$ ($m^2 s^{-1}$).

(ii) Using the equations indicated in (vi), Method 1, yields:

$(N_G)_{ME} = 2.13$ $(N'_L)_{ME} = 2.67$

$(N_G)_{MW} = 2.91$ $(N'_L)_{MW} = 3.08$

$(N_G)_{EW} = 2.65$ $(N'_L)_{EW} = 2.38$

(iii) From eqn. (10.5) $(\mathcal{N}_G)_{MM}=2.91$, $(\mathcal{N}_G)_{ME}=0.73$, $(\mathcal{N}_G)_{EM}=1.05\times 10^{-3}$, $(\mathcal{N}_G)_{EE}=2.16$.

(iv) From eqn. (10.5) for the liquid phase $(\mathcal{N}'_L)_{MM}=3.08$, $(\mathcal{N}'_L)_{ME}=0.34$, $(\mathcal{N}'_L)_{EM}=-8.03\times 10^{-4}$, $(\mathcal{N}'_L)_{EE}=2.62$.

(v) From eqn. (10.6) $m_{MM}=0.517$, $m_{ME}=79.34$, $m_{EM}=0.003\,48$, $m_{EE}=0.535$.

(vi) From eqn. (10.7) with $G=L$, $(\mathcal{N}_{OG})_{MM}=2.12$, $(\mathcal{N}_{OG})_{ME}=-95.6$, $(\mathcal{N}_{OG})_{EM}=-0.0032$, $(\mathcal{N}_{OG})_{EE}=1.65$.

(vii) Using eqn. (10.8) $G_{MM}=0.143$, $G_{ME}=15.5$, $G_{EM}=0.000\,52$, $G_{EE}=0.219$.

(viii) From eqn. (10.11) $\gamma=148.4$, and from eqn. (10.10)

for methanol, $E_{OG}=75\%$ so that $E_{MV}=92\%$

for ethanol, $E_{OG}=70\%$ so that $E_{MV}=86\%$

for water, $m_{WW}=0.517$, $E_{OG}=75\%$ and $E_{MV}=92\%$

10.6 Conclusion

On comparing the predicted efficiencies from each of the three methods, it is apparent that they are all quite similar with method 2 predicting slightly high values. The differences between the component efficiencies predicted by method 3, the most rigorous of the three methods, are not large.

There is an urgent need for tests of these multicomponent efficiency prediction methods against data from large-scale columns. This will indicate whether the simpler efficiency prediction methods will suffice or whether more complex methods are indeed necessary. Furthermore, it should be noted that multicomponent prediction methods are based on binary correlations, and so multicomponent efficiencies cannot be predicted any more accurately than we can predict binary efficiencies. As we have seen in earlier chapters, there is still room for improvement even for binaries, and binary efficiency prediction still remains the area where the biggest gains can be made from further research on distillation trays.

Appendix A

Horizontal momentum balance over the exit calming zone

Assuming a linear variation of froth density with height at ① in Fig. A1

$$\alpha_h = \frac{2\alpha(h_f - h)}{h_f} \tag{A1}$$

where α_h is the liquid holdup fraction at height h above the tray, and α is the mean liquid holdup fraction. From Newton's second law

$$\frac{dP}{dh} = -\alpha_h \rho_L g \tag{A2}$$

where P is the pressure at height h in the froth.

Integrating eqn. (A2) and setting $P = P_s$ at $h = h_f$ gives

$$P - P_s = \frac{2\alpha\rho_L g}{h_f}\left(\frac{h^2}{2} - h_f h + \frac{h_f^2}{2}\right) \tag{A3}$$

The total horizontal force F_x acting on the froth at section ① is

$$F_{x1} = \int_0^{h_f} (P - P_s) W \, dh \tag{A4}$$

Fig. A1. Control sections used.

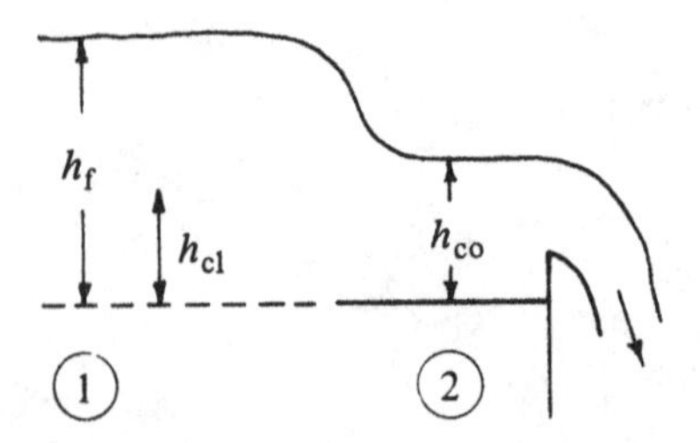

Substituting for eqn. (A3) and integrating gives

$$F_{x1} = \frac{\alpha \rho_L g W h_f^2}{3} \tag{A5}$$

At section ② the liquid is unaerated and has a density ρ_L so that

$$P - P_s = \rho_L g (h_{co} - h) \tag{A6}$$

and

$$F_{x2} = \frac{\rho_L g W h_{co}^2}{2}$$

A horizontal momentum balance on the liquid between ① and ② gives

$$\frac{Q_L^2 \rho_L}{h_{cl} W} + \frac{\alpha \rho_L g W h_f^2}{3} = \frac{Q_L^2 \rho_L}{h_{co} W} + \frac{h_{co}^2 W \rho_L g}{2} \tag{A7}$$

where the liquid velocity is assumed uniform over each section, and friction has been neglected.

Appendix B

Apparent Murphree vapour-phase tray efficiency in the presence of entrainment and weeping

e_0 = fraction of liquid entrained; β_0 = fraction of liquid weeping.

Calculated E^a_{MV} in the presence of entrainment. Lewis's case 1

$e_0 = 0$

Pe	λ_0 = 0.5	1.0	2.0	3.0
	$E_{OG} = 0.4$			
0	0.40	0.40	0.40	0.40
2	0.42	0.44	0.47	0.52
10	0.43	0.47	0.56	0.66
20	0.44	0.48	0.58	0.71
1000	0.44	0.49	0.61	0.77
	$E_{OG} = 0.6$			
0	0.60	0.60	0.60	0.60
2	0.64	0.68	0.77	0.87
10	0.68	0.77	0.99	1.28
50	0.70	0.81	1.12	1.57
1000	0.70	0.82	1.16	1.68
	$E_{OG} = 0.8$			
0	0.80	0.80	0.80	0.80
2	0.87	0.95	1.12	1.31
10	0.94	1.12	1.57	2.20
50	0.97	1.20	1.86	2.98
1000	0.98	1.22	1.97	3.32
	$E_{OG} = 1.0$			
0	1.00	1.00	1.00	1.00
2	1.11	1.24	1.51	1.84
10	1.23	1.52	2.32	3.55
50	1.28	1.67	2.94	5.38
1000	1.30	1.72	3.18	6.30

$e_0 = 0.1$

Pe	λ_0 = 0.5	1.0	2.0	3.0
	$E_{OG} = 0.4$			
0	0.39	0.39	0.38	0.38
2	0.41	0.42	0.45	0.48
10	0.43	0.46	0.53	0.61
20	0.43	0.47	0.55	0.65
1000	0.44	0.48	0.58	0.71
	$E_{OG} = 0.6$			
0	0.57	0.57	0.56	0.56
2	0.62	0.65	0.72	0.79
10	0.66	0.73	0.91	1.15
50	0.68	0.77	1.03	1.39
1000	0.68	0.78	1.06	1.48
	$E_{OG} = 0.8$			
0	0.75	0.74	0.74	0.73
2	0.82	0.88	1.01	1.17
10	0.90	1.04	1.40	1.91
50	0.93	1.12	1.66	2.53
1000	0.94	1.14	1.75	2.79
	$E_{OG} = 1.0$			
0	0.91	0.91	0.91	0.91
2	1.02	1.12	1.35	1.61
10	1.14	1.38	2.03	2.99
50	1.19	1.52	2.53	4.37
1000	1.21	1.56	2.73	5.04

$e_0 = 0.3$

Pe	λ_0 = 0.5	1.0	2.0	3.0
	$E_{OG} = 0.4$			
0	0.37	0.36	0.34	0.34
2	0.39	0.39	0.41	0.43
10	0.41	0.43	0.48	0.54
20	0.42	0.44	0.50	0.57
1000	0.42	0.45	0.53	0.62
	$E_{OG} = 0.6$			
0	0.53	0.51	0.49	0.48
2	0.58	0.59	0.63	0.68
10	0.62	0.67	0.79	0.95
50	0.64	0.71	0.89	1.13
1000	0.65	0.72	0.92	1.20
	$E_{OG} = 0.8$			
0	0.66	0.65	0.63	0.63
2	0.74	0.77	0.86	0.96
10	0.82	0.91	1.17	1.51
50	0.86	0.99	1.36	1.93
1000	0.87	1.00	1.43	2.10
	$E_{OG} = 1.0$			
0	0.77	0.77	0.77	0.77
2	0.88	0.95	1.11	1.29
10	0.99	1.17	1.62	2.24
50	1.05	1.28	1.97	3.11
1000	1.06	1.32	2.11	3.51

$e_0 = 0.5$

Pe	λ_0 = 0.5	1.0	2.0	3.0
	$E_{OG} = 0.4$			
0	0.35	0.33	0.31	0.30
2	0.38	0.37	0.37	0.38
10	0.40	0.41	0.44	0.48
20	0.41	0.42	0.46	0.51
1000	0.41	0.43	0.48	0.55
	$E_{OG} = 0.6$			
0	0.49	0.46	0.44	0.43
2	0.54	0.53	0.56	0.59
10	0.59	0.61	0.70	0.82
50	0.61	0.65	0.78	0.96
1000	0.61	0.66	0.81	1.01
	$E_{OG} = 0.8$			
0	0.59	0.57	0.56	0.55
2	0.67	0.69	0.75	0.82
10	0.75	0.82	1.00	1.24
50	0.79	0.88	1.16	1.55
1000	0.80	0.90	1.21	1.68
	$E_{OG} = 1.0$			
0	0.67	0.67	0.67	0.67
2	0.77	0.82	0.94	1.08
10	0.88	1.01	1.34	1.78
50	0.93	1.11	1.61	2.39
1000	0.95	1.14	1.71	2.65

Calculated E^a_{MV} in the presence of entrainment. Lewis's case 2

$e_0 = 0$

Pe	λ_0 = 0.5	1.0	2.0	3.0
	$E_{OG}=0.4$			
0	0.40	0.40	0.40	0.40
2	0.42	0.44	0.48	0.52
10	0.44	0.48	0.57	0.68
20	0.44	0.49	0.60	0.74
1000	0.45	0.50	0.63	0.81
	$E_{OG}=0.6$			
0	0.60	0.60	0.60	0.60
2	0.64	0.69	0.79	0.89
10	0.69	0.80	1.05	1.37
50	0.71	0.84	1.20	1.72
1000	0.71	0.86	1.25	1.85
	$E_{OG}=0.8$			
0	0.80	0.80	0.80	0.80
2	0.88	0.97	1.16	1.36
10	0.98	1.19	1.72	2.46
50	1.02	1.30	2.11	3.45
1000	1.03	1.33	2.25	3.89
	$E_{OG}=1.0$			
0	1.00	1.00	1.00	1.00
2	1.13	1.28	1.59	1.95
10	1.32	1.70	2.68	4.14
50	1.41	1.93	3.55	6.58
1000	1.43	2.00	3.90	7.82

$e_0 = 0.1$

Pe	λ_0 = 0.5	1.0	2.0	3.0
	$E_{OG}=0.4$			
0	0.39	0.38	0.38	0.38
2	0.41	0.42	0.45	0.49
10	0.43	0.46	0.54	0.64
20	0.43	0.48	0.57	0.68
1000	0.44	0.49	0.60	0.75
	$E_{OG}=0.6$			
0	0.57	0.57	0.56	0.56
2	0.62	0.65	0.73	0.81
10	0.67	0.76	0.97	1.23
50	0.69	0.81	1.10	1.52
1000	0.70	0.82	1.15	1.63
	$E_{OG}=0.8$			
0	0.75	0.74	0.74	0.73
2	0.83	0.90	1.05	1.22
10	0.94	1.11	1.55	2.13
50	0.98	1.22	1.88	2.93
1000	0.99	1.25	2.00	3.26
	$E_{OG}=1.0$			
0	0.91	0.91	0.91	0.91
2	1.04	1.16	1.42	1.71
10	1.23	1.54	2.34	3.48
50	1.32	1.75	3.06	5.34
1000	1.34	1.82	3.33	6.25

$e_0 = 0.3$

Pe	λ_0 = 0.5	1.0	2.0	3.0
	$E_{OG}=0.4$			
0	0.37	0.36	0.34	0.34
2	0.39	0.40	0.41	0.43
10	0.42	0.44	0.50	0.56
20	0.42	0.45	0.52	0.60
1000	0.43	0.46	0.55	0.66
	$E_{OG}=0.6$			
0	0.53	0.51	0.49	0.48
2	0.58	0.59	0.64	0.69
10	0.64	0.70	0.84	1.02
50	0.66	0.74	0.96	1.25
1000	0.67	0.76	0.99	1.33
	$E_{OG}=0.8$			
0	0.66	0.65	0.63	0.63
2	0.75	0.79	0.89	1.00
10	0.86	0.98	1.29	1.68
50	0.91	1.08	1.54	2.23
1000	0.92	1.11	1.63	2.45
	$E_{OG}=1.0$			
0	0.77	0.77	0.77	0.77
2	0.90	0.98	1.16	1.37
10	1.08	1.30	1.86	2.61
50	1.17	1.48	2.37	3.81
1000	1.19	1.54	2.57	4.35

$e_0 = 0.5$

Pe	λ_0 = 0.5	1.0	2.0	3.0
	$E_{OG}=0.4$			
0	0.35	0.33	0.31	0.30
2	0.38	0.37	0.38	0.39
10	0.41	0.42	0.46	0.50
20	0.41	0.43	0.48	0.54
1000	0.42	0.44	0.51	0.59
	$E_{OG}=0.6$			
0	0.49	0.46	0.44	0.43
2	0.54	0.54	0.57	0.61
10	0.61	0.64	0.75	0.88
50	0.63	0.69	0.85	1.06
1000	0.64	0.71	0.88	1.12
	$E_{OG}=0.8$			
0	0.59	0.57	0.56	0.55
2	0.68	0.70	0.77	0.85
10	0.80	0.88	1.10	1.39
50	0.85	0.97	1.31	1.79
1000	0.86	1.00	1.38	1.96
	$E_{OG}=1.0$			
0	0.67	0.67	0.67	0.67
2	0.79	0.85	0.99	1.14
10	0.96	1.13	1.54	2.07
50	1.04	1.28	1.93	2.91
1000	1.07	1.33	2.08	3.27

Calculated E^a_{MV} in the presence of entrainment. Lewis's case 3

$e_0=0$				
	λ_0			
	0.5	1.0	2.0	3.0
Pe	$E_{OG}=0.4$			
0	0.40	0.40	0.40	0.40
2	0.42	0.44	0.47	0.51
10	0.43	0.47	0.55	0.65
20	0.44	0.48	0.57	0.69
1000	0.44	0.49	0.60	0.75
	$E_{OG}=0.6$			
0	0.60	0.60	0.60	0.60
2	0.64	0.68	0.76	0.85
10	0.68	0.76	0.96	1.20
50	0.69	0.80	1.06	1.44
1000	0.70	0.81	1.10	1.53
	$E_{OG}=0.8$			
0	0.80	0.80	0.80	0.80
2	0.87	0.94	1.09	1.26
10	0.93	1.07	1.44	1.94
50	0.96	1.14	1.65	2.49
1000	0.96	1.16	1.71	2.72
	$E_{OG}=1.0$			
0	1.00	1.00	1.00	1.00
2	1.10	1.20	1.44	1.72
10	1.17	1.39	1.98	2.89
50	1.20	1.47	2.33	3.95
1000	1.21	1.50	2.46	4.43

$e_0=0.1$				
	λ_0			
	0.5	1.0	2.0	3.0
Pe	$E_{OG}=0.4$			
0	0.39	0.38	0.38	0.38
2	0.41	0.42	0.45	0.48
10	0.43	0.45	0.52	0.60
20	0.43	0.46	0.54	0.64
1000	0.43	0.47	0.57	0.69
	$E_{OG}=0.6$			
0	0.57	0.57	0.56	0.56
2	0.62	0.64	0.71	0.78
10	0.65	0.72	0.88	1.08
50	0.67	0.75	0.97	1.27
1000	0.68	0.77	1.00	1.35
	$E_{OG}=0.8$			
0	0.75	0.74	0.74	0.73
2	0.82	0.87	0.99	1.12
10	0.88	1.00	1.29	1.69
50	0.91	1.06	1.47	2.12
1000	0.92	1.08	1.54	2.30
	$E_{OG}=1.0$			
0	0.91	0.91	0.91	0.91
2	1.00	1.09	1.29	1.51
10	1.08	1.26	1.74	2.45
50	1.11	1.34	2.03	3.25
1000	1.12	1.36	2.13	3.59

$e_0=0.3$				
	λ_0			
	0.5	1.0	2.0	3.0
Pe	$E_{OG}=0.4$			
0	0.37	0.36	0.34	0.34
2	0.39	0.39	0.40	0.42
10	0.41	0.43	0.47	0.52
20	0.41	0.44	0.49	0.55
1000	0.42	0.45	0.51	0.60
	$E_{OG}=0.6$			
0	0.53	0.51	0.49	0.48
2	0.57	0.58	0.62	0.66
10	0.61	0.65	0.76	0.89
50	0.63	0.68	0.83	1.04
1000	0.63	0.69	0.86	1.09
	$E_{OG}=0.8$			
0	0.66	0.65	0.63	0.63
2	0.73	0.76	0.84	0.93
10	0.80	0.87	1.07	1.34
50	0.82	0.92	1.21	1.63
1000	0.83	0.94	1.26	1.75
	$E_{OG}=1.0$			
0	0.77	0.77	0.77	0.77
2	0.86	0.92	1.06	1.22
10	0.94	1.07	1.40	1.86
50	0.97	1.13	1.60	2.36
1000	0.98	1.15	1.67	2.57

$e^0=0.5$				
	λ_0			
	0.5	1.0	2.0	3.0
Pe	$E_{OG}=0.4$			
0	0.35	0.33	0.31	0.30
2	0.37	0.37	0.37	0.38
10	0.39	0.40	0.43	0.46
20	0.40	0.41	0.45	0.49
1000	0.41	0.42	0.47	0.53
	$E_{OG}=0.6$			
0	0.49	0.46	0.44	0.43
2	0.53	0.53	0.55	0.58
10	0.57	0.60	0.67	0.77
50	0.59	0.63	0.73	0.88
1000	0.60	0.64	0.75	0.92
	$E_{OG}=0.8$			
0	0.59	0.57	0.56	0.55
2	0.66	0.67	0.73	0.79
10	0.72	0.77	0.92	1.11
50	0.75	0.82	1.03	1.33
1000	0.76	0.84	1.07	1.41
	$E_{OG}=1.0$			
0	0.67	0.67	0.67	0.67
2	0.75	0.80	0.90	1.02
10	0.83	0.92	1.17	1.50
50	0.86	0.98	1.32	1.84
1000	0.87	1.00	1.38	1.98

Calculated E^{a}_{MV} in the presence of weeping. Lewis's case 1

$\beta_o=0$				
	λ_o			
	0.5	1.0	2.0	3.0
Pe	$E_{OG}=0.4$			
0	0.40	0.40	0.40	0.40
2	0.42	0.44	0.48	0.52
10	0.43	0.47	0.56	0.66
20	0.44	0.48	0.58	0.71
1000	0.44	0.49	0.61	0.77
	$E_{OG}=0.6$			
0	0.60	0.60	0.60	0.60
2	0.64	0.68	0.77	0.87
10	0.68	0.77	0.99	1.28
50	0.70	0.81	1.12	1.57
1000	0.70	0.82	1.16	1.68
	$E_{OG}=0.8$			
0	0.80	0.80	0.80	0.80
2	0.87	0.95	1.12	1.31
10	0.94	1.12	1.57	2.20
50	0.97	1.20	1.86	2.98
1000	0.98	1.22	1.97	3.32
	$E_{OG}=1.0$			
0	1.00	1.00	1.00	1.00
2	1.11	1.24	1.51	1.84
10	1.23	1.52	2.32	3.55
50	1.28	1.67	2.94	5.38
1000	1.30	1.72	3.18	6.30

$\beta_o=0.1$				
	λ_o			
	0.5	1.0	2.0	3.0
Pe	$E_{OG}=0.4$			
0	0.40	0.40	0.40	0.40
2	0.42	0.43	0.47	0.50
10	0.43	0.46	0.53	0.58
20	0.44	0.47	0.54	0.61
1000	0.44	0.48	0.56	0.64
	$E_{OG}=0.6$			
0	0.60	0.60	0.60	0.60
2	0.63	0.67	0.74	0.81
10	0.67	0.74	0.88	1.00
50	0.68	0.77	0.95	1.10
1000	0.68	0.78	0.97	1.12
	$E_{OG}=0.8$			
0	0.80	0.80	0.80	0.80
2	0.86	0.93	1.05	1.16
10	0.92	1.06	1.30	1.48
50	0.94	1.12	1.42	1.61
1000	0.95	1.14	1.46	1.64
	$E_{OG}=1.0$			
0	1.00	1.00	1.00	1.00
2	1.09	1.20	1.39	1.55
10	1.18	1.41	1.77	1.98
50	1.22	1.51	1.95	2.11
1000	1.23	1.55	2.00	2.14

$\beta_o=0.3$				
	λ_o			
	0.5	1.0	2.0	3.0
Pe	$E_{OG}=0.4$			
0	0.40	0.40	0.40	0.40
2	0.41	0.42	0.44	0.46
10	0.42	0.44	0.47	0.49
20	0.42	0.44	0.48	0.50
1000	0.43	0.45	0.48	0.50
	$E_{OG}=0.6$			
0	0.60	0.60	0.60	0.60
2	0.63	0.65	0.69	0.71
10	0.65	0.69	0.74	0.77
50	0.66	0.70	0.76	0.77
1000	0.66	0.71	0.77	0.78
	$E_{OG}=0.8$			
0	0.80	0.80	0.80	0.80
2	0.84	0.88	0.94	0.98
10	0.89	0.95	1.03	1.05
50	0.90	0.98	1.05	1.04
1000	0.91	0.99	1.05	1.04
	$E_{OG}=1.0$			
0	1.00	1.00	1.00	1.00
2	1.07	1.13	1.21	1.26
10	1.13	1.23	1.32	1.31
50	1.16	1.28	1.34	1.27
1000	1.17	1.29	1.34	1.25

$\beta_o=0.5$				
	λ_o			
	0.5	1.0	1.5	2.0
Pe	$E_{OG}=0.4$			
0	0.40	0.40	0.40	0.40
2	0.41	0.41	0.42	0.42
10	0.41	0.42	0.43	0.43
20	0.41	0.42	0.43	0.43
1000	0.41	0.42	0.43	0.43
	$E_{OG}=0.6$			
0	0.60	0.60	0.60	0.60
2	0.62	0.63	0.64	0.64
10	0.63	0.64	0.65	0.66
50	0.63	0.65	0.66	0.66
1000	0.63	0.65	0.66	0.65
	$E_{OG}=0.8$			
0	0.80	0.80	0.80	0.80
2	0.83	0.85	0.86	0.87
10	0.85	0.87	0.88	0.88
50	0.86	0.88	0.88	0.87
1000	0.86	0.88	0.88	0.79
	$E_{OG}=1.0$			
0	1.00	1.00	1.00	1.00
2	1.04	1.07	1.09	1.10
10	1.08	1.11	1.11	1.11
50	1.09	1.12	1.11	1.08
1000	1.09	1.12	1.10	1.03

Calculated E^a_{MV} in the presence of weeping. Lewis's case 2

$\beta_0=0$

	λ_0			
	0.5	1.0	2.0	3.0
Pe	$E_{OG}=0.4$			
0	0.40	0.40	0.40	0.40
2	0.42	0.44	0.48	0.52
10	0.44	0.48	0.57	0.68
20	0.44	0.49	0.60	0.74
1000	0.45	0.50	0.63	0.81
	$E_{OG}=0.6$			
0	0.60	0.60	0.60	0.60
2	0.64	0.69	0.79	0.89
10	0.69	0.80	1.05	1.37
50	0.71	0.84	1.20	1.72
1000	0.71	0.86	1.25	1.85
	$E_{OG}=0.8$			
0	0.80	0.80	0.80	0.80
2	0.88	0.97	1.15	1.36
10	0.98	1.19	1.73	2.46
50	1.02	1.30	2.11	3.45
1000	1.03	1.33	2.25	3.89
	$E_{OG}=1.0$			
0	1.00	1.00	1.00	1.00
2	1.13	1.28	1.59	1.95
10	1.32	1.70	2.68	4.14
50	1.41	1.93	3.55	6.58
1000	1.43	2.00	3.90	7.82

$\beta_0=0.1$

	λ_0			
	0.5	1.0	2.0	3.0
Pe	$E_{OG}=0.4$			
0	0.40	0.40	0.40	0.40
2	0.42	0.44	0.47	0.51
10	0.43	0.47	0.55	0.63
20	0.44	0.49	0.58	0.66
1000	0.44	0.49	0.59	0.71
	$E_{OG}=0.6$			
0	0.60	0.60	0.60	0.60
2	0.64	0.68	0.76	0.85
10	0.68	0.77	0.96	1.16
50	0.70	0.81	1.06	1.34
1000	0.70	0.82	1.10	1.40
	$E_{OG}=0.8$			
0	0.80	0.80	0.80	0.80
2	0.87	0.95	1.10	1.25
10	0.96	1.14	1.50	1.88
50	1.00	1.23	1.73	2.26
1000	1.00	1.25	1.80	2.38
	$E_{OG}=1.0$			
0	1.00	1.00	1.00	1.00
2	1.12	1.24	1.49	1.73
10	1.29	1.58	2.19	2.79
50	1.37	1.76	2.61	3.44
1000	1.39	1.82	2.75	3.65

$\beta_0=0.3$

	λ_0			
	0.5	1.0	2.0	3.0
Pe	$E_{OG}=0.4$			
0	0.40	0.40	0.40	0.40
2	0.41	0.43	0.45	0.47
10	0.43	0.45	0.50	0.55
20	0.43	0.46	0.52	0.56
1000	0.43	0.47	0.53	0.59
	$E_{OG}=0.6$			
0	0.60	0.60	0.60	0.60
2	0.63	0.66	0.72	0.77
10	0.66	0.73	0.83	0.93
50	0.68	0.75	0.89	1.00
1000	0.68	0.76	0.90	1.02
	$E_{OG}=0.8$			
0	0.80	0.80	0.80	0.80
2	0.86	0.91	1.01	1.09
10	0.93	1.04	1.23	1.37
50	0.95	1.10	1.32	1.50
1000	0.96	1.11	1.35	1.53
	$E_{OG}=1.0$			
0	1.00	1.00	1.00	1.00
2	1.09	1.17	1.33	1.45
10	1.23	1.40	1.68	1.87
50	1.29	1.51	1.83	2.06
1000	1.30	1.54	1.88	2.11

$\beta_0=0.5$

	λ_0			
	0.5	1.0	1.5	2.0
Pe	$E_{OG}=0.4$			
0	0.40	0.40	0.40	0.40
2	0.41	0.42	0.43	0.43
10	0.42	0.44	0.45	0.47
20	0.42	0.44	0.46	0.47
1000	0.42	0.44	0.46	0.48
	$E_{OG}=0.6$			
0	0.60	0.60	0.60	0.60
2	0.62	0.64	0.66	0.68
10	0.64	0.68	0.72	0.75
50	0.65	0.70	0.74	0.77
1000	0.66	0.71	0.75	0.78
	$E_{OG}=0.8$			
0	0.80	0.80	0.80	0.80
2	0.84	0.88	0.91	0.93
10	0.89	0.96	1.01	1.05
50	0.91	0.99	1.05	1.10
1000	0.91	1.00	1.06	1.11
	$E_{OG}=1.0$			
0	1.00	1.00	1.00	1.00
2	1.07	1.12	1.17	1.21
10	1.14	1.26	1.33	1.39
50	1.18	1.32	1.40	1.46
1000	1.19	1.33	1.42	1.48

Calculated E^{a}_{MV} in the presence of weeping. Lewis's case 3

$\beta_0=0$				
	λ_0			
	0.5	1.0	2.0	3.0
Pe	$E_{OG}=0.4$			
0	0.40	0.40	0.40	0.40
2	0.42	0.44	0.47	0.51
10	0.43	0.47	0.55	0.65
20	0.44	0.48	0.57	0.69
1000	0.44	0.49	0.60	0.75
	$E_{OG}=0.6$			
0	0.60	0.60	0.60	0.60
2	0.64	0.68	0.76	0.85
10	0.68	0.76	0.95	1.20
50	0.69	0.80	1.06	1.44
1000	0.70	0.81	1.10	1.53
	$E_{OG}=0.8$			
0	0.80	0.80	0.80	0.80
2	0.87	0.94	1.09	1.26
10	0.93	1.07	1.44	1.94
50	0.96	1.14	1.65	2.49
1000	0.97	1.16	1.73	2.72
	$E_{OG}=1.0$			
0	1.00	1.00	1.00	1.00
2	1.10	1.20	1.44	1.72
10	1.17	1.39	1.98	2.89
50	1.20	1.47	2.33	3.95
1000	1.21	1.50	2.46	4.43

$\beta_0=0.1$				
	λ_0			
	0.5	1.0	2.0	3.0
Pe	$E_{OG}=0.4$			
0	0.40	0.40	0.40	0.40
2	0.42	0.43	0.46	0.49
10	0.43	0.46	0.52	0.57
20	0.43	0.47	0.53	0.60
1000	0.44	0.48	0.55	0.62
	$E_{OG}=0.6$			
0	0.60	0.60	0.60	0.60
2	0.63	0.67	0.73	0.79
10	0.67	0.73	0.85	0.96
50	0.68	0.76	0.91	1.03
1000	0.68	0.77	0.93	1.05
	$E_{OG}=0.8$			
0	0.80	0.80	0.80	0.80
2	0.86	0.92	1.02	1.12
10	0.91	1.02	1.21	1.36
50	0.93	1.07	1.29	1.43
1000	0.94	1.08	1.32	1.44
	$E_{OG}=1.0$			
0	1.00	1.00	1.00	1.00
2	1.08	1.17	1.33	1.48
10	1.14	1.30	1.57	1.74
50	1.17	1.35	1.65	1.75
1000	1.17	1.36	1.64	1.72

$\beta_0=0.3$				
	λ_0			
	0.5	1.0	2.0	3.0
Pe	$E_{OG}=0.4$			
0	0.40	0.40	0.40	0.40
2	0.41	0.42	0.44	0.45
10	0.42	0.44	0.46	0.48
20	0.42	0.44	0.47	0.49
1000	0.43	0.45	0.48	0.49
	$E_{OG}=0.6$			
0	0.60	0.60	0.60	0.60
2	0.62	0.65	0.68	0.71
10	0.64	0.68	0.72	0.74
50	0.65	0.69	0.73	0.74
1000	0.66	0.70	0.74	0.73
	$E_{OG}=0.8$			
0	0.80	0.80	0.80	0.80
2	0.84	0.88	0.93	0.96
10	0.87	0.92	0.97	0.98
50	0.89	0.94	0.97	0.93
1000	0.89	0.94	0.96	0.91
	$E_{OG}=1.0$			
0	1.00	1.00	1.00	1.00
2	1.06	1.11	1.18	1.22
10	1.09	1.15	1.20	1.18
50	1.10	1.16	1.15	1.06
1000	1.10	1.15	1.14	1.00

$\beta_0=0.5$				
	λ_0			
	0.5	1.0	1.5	2.0
Pe	$E_{OG}=0.4$			
0	0.40	0.40	0.40	0.40
2	0.41	0.41	0.42	0.42
10	0.41	0.42	0.42	0.43
20	0.41	0.42	0.43	0.43
1000	0.41	0.42	0.43	0.43
	$E_{OG}=0.6$			
0	0.60	0.60	0.60	0.60
2	0.62	0.63	0.63	0.64
10	0.62	0.64	0.64	0.64
50	0.63	0.64	0.64	0.63
1000	0.63	0.64	0.63	0.62
	$E_{OG}=0.8$			
0	0.80	0.80	0.80	0.80
2	0.83	0.84	0.85	0.86
10	0.84	0.85	0.85	0.84
50	0.84	0.84	0.82	0.79
1000	0.84	0.84	0.82	0.76
	$E_{OG}=1.0$			
0	1.00	1.00	1.00	1.00
2	1.03	1.05	1.07	1.08
10	1.04	1.05	1.04	1.02
50	1.03	1.02	0.98	0.92
1000	1.03	1.00	0.95	0.86

References

Aerov, M. E., Bystrova, T. A. and Koltunova, L. N. (1970). Mass transfer in the gas phase on bubble plates without overflow devices, *Teoreticheskie Osnovy Khimicheskoi Tekhnologii* **4**, (4), 467

Aiba, S. and Yamada, T. (1959). Studies on entrainment, *AIChE J.* **5**, (4), 506

AIChE (1958). *Bubble Tray Design Manual.* AIChE, New York

AIChE (1986). *Tray Distillation Columns – A Guide to Performance Evaluation.* AIChE, New York

Aittamaa, J. (1981). Estimating multicomponent plate efficiencies in distillation, *Kemia-Kemi* **8**, (5), 295

Akselrod, L. S. and Yusova, G. M. (1957). Dispersity of the liquid in the interplate space in bubble towers. *J. Appl. Chem. USSR* **30**, 739

Aleksandrov, I. A. and Shoblo, A. (1960). Entrainment of a liquid by a gas with screen type plates, *Khim. Tekhnol. Topl. Masel.* **5**, (9), 42

Aleksandrov, I. A. and Shoblo, A. (1962). Determination of the amount of interplate entrainment of liquid in rectification columns, *Khim. Tekhnol. Topl. Masel.* **7**, (8), 53

Al Taweel, A. M., Divakarla, R. and Gomaa, H. G. (1984). Measurement of large gas–liquid interfacial areas, *Can. J. Chem. Engng.* **62**, (Feb.), 73

Andrew, S. P. S. (1960). Frothing in two component liquid mixtures. Int. Symp. on Distillation, Inst. Chem. Engrs., Brighton, p. 73

Andrew, S. P. S. (1961). Aspects of gas–liquid mass transfer. *Alta Tecnologia Chimica*, p. 153

Andrew, S. P. S. (1969). Hydrodynamics of sieve plates at high liquid and vapour throughputs, *Inst. Chem. Engrs. Symp. Series No. 32*, p. 2:49

Arnold, D. S., Plank, C. A. and Schoenborn, E. M. (1952). Performance of perforated plate distillation columns, *Chem. Eng. Prog.* **48**, (12), 633

Asano, K. and Fujita, S. (1966). Vapour and liquid phase mass transfer coefficients in tray towers, *Kagaku Kogaku* **4**, 330 & 369

Ashley, M. J. and Haselden, G. G. (1970). The calculation of plate efficiency under conditions of finite mixing in both phases in multiplate columns, and the potential advantage of parallel flow, *Chem. Engng. Sci.* **25**, 1665

Ashley, M. J. and Haselden, G. G. (1972). Effectiveness of vapour–liquid contacting on a sieve plate, *Trans. Inst. Chem. Engrs.* **50**, 119

Assenov, A., Elenkov, D. and Vulchev, L. (1969). Effect of flow rate on entrainment of droplets above sieve plates, when employed liquids contain surface-active admixtures, *Comptes rendus de l'Academie bulgare des Sciences* **22**, (8), 879

Azbel, D. S. (1963). The hydrodynamics of bubbler processes, *Int. Chem. Engng.* **3**, (3), 319

Backhurst, J. R. and Harker, J. H. (1973). *Process Plant Design.* Heinemann

Bain, J. F. and Van Winkle, M. (1961). A study of entrainment, perforated plate column, air–water system, *AIChE J.* **7**, (3), 363
Bainbridge, G. S. and Sawistowski, H. (1964). Surface tension effects in sieve plate distillation columns, *Chem. Engng. Sci.* **19**, 992
Banerjee, T. S., Roy, N. K. and Rao, M. N. (1969*a*). Studies on entrainment of drops in flow through orifices: Part I – Mechanism of entrainment and drop size distribution, *Ind. J. Techn.* **7**, (Oct.), 301
Banerjee, T. S., Roy, N. K. and Rao, M. N. (1969*b*). Studies on entrainment of drops in flow through orifices: Part II – Magnitude of entrainment from liquids, *Ind. J. Techn.* **7** (Oct.), 308
Banik, S. (1982). PhD Thesis, University of Manchester Institute of Science and Technology
Barber, A. D. and Hartland, S. (1975). The collapse of cellular foams, *Trans. Inst. Chem. Engrs.* **53**, 106
Barber, A. D. and Wijn, E. F. (1979). Foaming in crude distillation units, *Inst. Chem. Engrs. Symp. Series No. 56*, p. 3.1/15
Barker, P. E. and Choudhury, M. H. (1959). Performance of bubble cap trays, *Brit. Chem. Engng.* **4**, (6), 348
Barker, P. E. and Self, M. F. (1962). The evaluation of liquid mixing effects on a sieve plate using unsteady and steady state tracer techniques, *Chem. Engng. Sci.* **17**, 541
Bartholomai, G. B., Gardner, R. G. and Hamilton, W. (1972). Characteristics of a sieve plate with downcomer, *Brit. Chem. Eng. & Proc. Tech.* **17**, (1), 48
Bell, R. L. (1972). Residence time and fluid mixing on commercial scale sieve trays, *AIChE J.* **18**, (3), 498
Bell, R. L. and Solari, R. B. (1974). Effect of nonuniform velocity fields and retrograde flow on distillation tray efficiency, *AIChE J.* **20**, (4), 688
Bene, T. (1964). Determination of local resistance coefficients for perforated plates, *Int. Chem. Engng.* **4**, (4), 625
Bennett, D. L., Agrawal, R. and Cook, P. J. (1983). New pressure drop correlation for sieve tray distillation columns, *AIChE J*, **29**, (3), 434
Bentham, J. B. and Darton, R. C. (1983). Capacity and flow regimes on distillation trays at high pressure. EFChE Dist. Abs. and Extn. Working Party Meeting, June, Winterthur, Switzerland
Bernard, J. D. T. and Sargent, R. W. H. (1966). The hydrodynamic performance of a sieve plate distillation column, *Trans. Inst. Chem. Engrs.* **44**, T314
Biddulph, M. W. (1975*a*). Oscillating behaviour on distillation trays – II, *AIChE J.* **21**, (1), 41
Biddulph, M. W. (1975*b*). Multicomponent distillation simulation – distillation of air, *AIChE J.* **21**, (2), 327
Biddulph, M. W. (1977). Predicted comparisons of the efficiency of large valve trays and large sieve trays, *AIChE J.* **23**, (5), 770
Biddulph, M. W. and Ashton, N. (1977). Deducing multicomponent distillation efficiencies from industrial data, *Chem. Engng. J.* **14**, 7
Biddulph, M. W. and Stephens, D. J. (1974). Oscillating behaviour on distillation trays, *AIChE J.* **20**, (1), 60
Bikerman, J. J. (1973). *Foams*. Springer-Verlag, New York
Billet, R. (1979). *Distillation Engineering*. Heyden
Billet, R., Conrad, S. and Grubb, C. M. (1969). Some aspects of the choice of distillation equipment, *Inst. Chem. Engrs. Symp. Series No. 32*, p. 5:111
Bird, R. B., Stewart, W. E. and Lightfoot, E. N. (1960). *Transport Phenomena*. John Wiley & Sons
Bolles, W. L. (1946). Rapid graphical method of estimating tower diameter and tray spacing of bubble-plate fractionators, *Pet. Refiner*, **25**, (12), 104

Bolles, W. L. (1956). Optimum bubble-cap tray design, *Petroleum Processing* **11**, (2), 65
Bolles, W. L. (1963). In Smith, B. D. *Design of Equilibrium Stage Processes*, Chap. 14. McGraw-Hill
Bolles, W. L. (1967). The solution of a foam problem, *Chem. Engng. Prog.* **63**, (9), 48
Bolles, W. L. (1976*a*). Estimating valve tray performance, *Chem. Eng. Prog.* **72**, (9), 43
Bolles, W. L. (1976*b*). Multipass flow distribution and mass transfer efficiency for distillation plates, *AIChE J.* **22**, (1), 153
Brambilla, A. (1976). The effect of vapour mixing on efficiency of large diameter distillation plates, *Chem. Engng. Sci.* **31**, 517
Brambilla, A. and Nencetti, G. F. (1973). Pressure drop for rectangular valve trays, *Ing. Chim. Ital.* **9**, (5), 67
Brambilla, A., Gianolio, E. and Nencetti, G. F. (1979). Fluodynamic behaviour and efficiency of sieve trays at low gas rates, *Inst. Chem. Engrs. Symp. Series No. 56*, p. 3.2/1
Brambilla, A., Nardini, G., Nencetti, G. F. and Zanelli, S. (1969). Hydrodynamic behaviour of distillation columns, *Inst. Chem. Engrs. Symp. Series No. 32*, p. 2:63
Brierley, R. J. P., Whyman, P. J. M. and Erskine, J. B. (1979). Flow induced vibration of distillation and absorption column trays, *Inst. Chem. Engrs. Symp. Series No. 56*, p. 2.4/45
Brown, R. S. (1958). PhD Thesis, University of California, Berkeley
Brumbaugh, K. H. and Berg, J. C. (1973). Effect of surface active agents on a sieve plate distillation column, *AIChE J.* **19**, (5), 1078
Burgess, J. M. and Calderbank, P. H. (1975). The measurement of bubble parameters in two phase dispersions, *Chem. Engng. Sci.* **30**, 743 & 1107
Calcaterra, R. J., Nicholls, C. W. and Weber, J. H. (1968). Free and captured entrainment and plate spacing in a perforated tray column, *Brit. Chem. Engng.* **13**, (9), 1294
Calderbank, P. H. (1956). Gas–liquid contacting on plates, *Trans. Inst. Chem. Engrs.* **34**, 79
Calderbank, P. H. (1959). Physical rate processes in industrial fermentation – Part II, *Trans. Inst. Chem. Engrs.* **37**, 173
Calderbank, P. H. and Moo-Young, M. B. (1960). The mass transfer efficiency of distillation and gas-absorption plate columns – Part II. Int. Symp. on Distillation, Inst. Chem. Engrs., Brighton, p. 59
Calderbank, P. H. and Pereira, J. (1977). The prediction of distillation plate efficiencies from froth properties, *Chem. Engng. Sci.* **32**, 1427
Calderbank, P. H. and Pereira, J. (1979). A micro-processor system to monitor distillation plate efficiencies, *Inst. Chem. Engrs. Symp. Series No. 56*, p. 2.3/27
Calderbank, P. H. and Rennie, J. (1962). The physical properties of foams and froths formed on sieve plates, *Trans. Inst. Chem. Engrs.* **40**, 3
Canfield, F. B. (1984). Computer simulation of the parastillation process, *Chem. Engng. Prog.* **80**, (2), 58
Carr, J. (1972). *The Second Oldest Profession – An Informal History of Moonshining in America*. Prentice Hall
Cervenka, J. and Kolar, V. (1973*a*). The structure and height of the gaseous–liquid mixture on sieve plates without downcomers, *Chem. Engng. J.* **6**, 45
Cervenka, J. and Kolar, V. (1973*b*). Hydrodynamics of plate volumns VIII. *Czech. Chem. Comm.* **38**, 2891
Cervenka, J. and Kolar, V. (1974). The pressure drop of sieve-plate columns, *Czech. Chem. Comm.* **39**, 2733
Chan, H. and Fair, J. R. (1984). Prediction of point efficiencies on sieve trays, *Ind. Eng. Chem. Proc. Des. Dev.* **23**, 814 & 820
Chase, J. D. (1967). Sieve tray design, *Chem. Engng.* **74**, (16), 105
Chekov, O. S., Kutepov, A. M., Suleimenov, M. K., and Kuchmistyi, B. I. (1975). Calculation of interplate entrainment and liquid-phase mass transfer on film plates, *Teor. Osn. Khim. Teknol.* **9** (1), 20

Cheng, S. I. and Teller, A. J. (1961). Free entrainment behaviour on sieve trays, *AIChE J.* 7, (2), 282

Chu, Ju. Chin, Donovan, J. R., Bosewell, B. C. and Furmeister, L. C. (1951). Plate efficiency correlation in distilling columns and gas absorbers, *J. Appl. Chem.* **1**, (Dec.), 529

Cicalese, J. J., Davies, J. A. Harrington, P. J., Houghland, G. S., Hutchinson, A. J. L. and Walsh, T. J. (1947*a*). Miscellaneous fractionating techniques, *Petrol. Refiner* **26**, (4), 431

Cicalese, J. J., Davies, J. A., Harrington, P. J., Houghland, G. S., Hutchinson, A. J. L. and Walsh, T. J. (1947*b*). Appendix: Miscellaneous fractionating techniques. Study of alkylation – plant isobutane tower performance, *Petrol. Refiner* **26**, (5), 495

Clift, R., Grace, J. R. and Weber, M. E. (1978). *Bubble Drops and Particles.* Academic Press

Colburn, A. P. (1936). Effect of entrainment on plate efficiency in distillation, *Ind. Eng. Chem.* **28**, 526

Colwell, C. J. (1979). Clear liquid height and froth density on sieve trays, *Ind. Eng. Chem. Proc. Des. Dev.* **20**, (2), 298

Crozier, R. D. (1956). PhD Thesis, University of Michigan

D'arcy, D. (1975). Bubbling from submerged orifices. AIChE meeting, 8 Sept., Boston (Atomic Energy of Canada Report 5257)

Danckwerts, P. V. (1953). Continuous flow systems. Distribution of residence times, *Chem. Engng. Sci.* **2**, 1

Danckwerts, P. V. (1970). *Gas–Liquid Reactions.* McGraw-Hill, New York

Danckwerts, P. V. and Sharma, M. M. (1966). The absorption of carbon dioxide into solutions of alkalis and amines, *The Chem. Engr.* (Oct.), CE244

Danckwerts, P. V., Smith, W. V. and Sawistowski, H. (1960). The effects of heat transfer and interfacial tension in distillation. Int. Symp. on Distillation, Inst. Chem. Engrs., Brighton, p. 7

Darton, R. C. (1979). Discussion volume, *Inst. Chem. Engrs. Symp. Series No. 56*, 28

Davidson, J. F. and Schuler, B. O. G. (1960*a*). Bubble formation at an orifice in a viscous liquid, *Trans. Inst. Chem. Engrs.* **38**, 144

Davidson, J. F. and Schuler, B. O. G. (1960*b*). Bubble formation at an orifice in an inviscid liquid, *Trans. Inst. Chem. Engrs.* **38**, 335

Davies, B. T. and Porter, K. E. (1965). Some observations of sieve tray froths. Proc. Symposium on Two Phase Flow, 21–23 June, University of Exeter, UK, p. F301

Davies, J. A. (1947). Bubble tray hydraulics, *Ind. Engng. Chem.* **39**, (6), 774

Davy, C. A. E. and Haselden, G. G. (1975). Prediction of the pressure drop across sieve trays, *AIChE J*, **21**, (6), 1218

Delnicki, W. V. and Wagner, J. L. (1970). Performance of multiple downcomer trays, *Chem. Eng. Prog.* **66**, (3), 50

Dhulesia, H. (1980). PhD Thesis, University of Manchester Institute of Science and Technology

Dhulesia, H. (1983). Operating flow regimes on the valve tray, *Chem. Eng. Res. Des.* **61**, 329

Dhulesia, H. (1984). Clear liquid height on sieve and valve trays, *Chem. Eng. Res. Des.* **62**, 321

Diener, D. A. (1967). Calculation of effect of vapour mixing on tray efficiency, *Ind. Eng. Chem. Proc. Des. Dev.* **6**, (4), 499

Diener, D. A. and Gerster, J. A. (1968). Point efficiencies in distillation of acetone–methanol–water, *Ind. Eng. Chem. Proc. Des. Dev.* **7**, (3), 339

Doig, I. D. (1971). Variation of the operating pressure to manipulate distillation processes, *Australian Chem. Engng.* **12**, (6), 5

Drickamer, H. G. and Bradford, J. R. (1943). Overall plate efficiency of commercial hydrocarbon fractionating columns as a function of viscosity, *Trans. Am. Inst. Chem. Engrs.* **39**, 319

Drogaris, G. and Lockett, M. J. (1979). The deleterious effect of vapour entrainment under the downcomer determined by Standart's entrainment model, *Chem. Eng. Comm.* **3**, 291

Dytnerskii, Y. I. and Andreev, V. I. (1966). Mechanism of spray entrainment in plate columns, *Tr. Mosk. Khim-Tekhnol. ivst.* **51**, 27

Dytnerskii, Y. I., Kasatkim, A. G., Kochergin, N. V. and Gervits, V. M. (1965). Calculating the hydraulics and mass transfer on valve plates, *Int. Chem. Eng.* **5**, (1), 95

Eben, C. D. and Pigford, R. L. (1965). Gas absorption with chemical reaction on a sieve tray, *Chem. Engng. Sci.* **20**, 803

Economopoulos, A. P. (1978). Computer design of sieve trays and tray columns, *Chem. Engng.* **85**, (27), 109

Eduljee, H. E. (1958*a*). Design of sieve-type distillation plates, *Brit. Chem. Engng.* **3**, (1), 14

Eduljee, H. E. (1958*b*). Entrainment from bubble cap distillation plates, *Brit. Chem. Engng.* **3**, (9), 474

Eduljee, H. E. (1965). Design of bubble cap distillation plates, *Brit. Chem. Eng.* **10**, (2), 103

Eduljee, H. E. (1972). Sieve plates – minimum vapour velocity, *The Chem. Engr.* (Mar.), 123

Elenkov, D. and Vulchev, L. (1968). Effect of soluble surface active agents on entrainment of liquid in sieve plate columns, *Comptes rendus de l'Academie bulgare des Sciences*, **21**, (6), 545

Ellis, S. R. M. and Biddulph, M. W. (1967). The effect of surface tension characteristics on plate efficiencies, *Trans. Inst. Chem. Engrs.* **45**, T223

Ellis, S. R. M. and Hardwick, M. J. (1969). Effect of reflux on plate efficiency, *Inst. Chem. Engrs. Symp. Series No. 32*, p. 1:29

Ellis, S. R. M. and Shelton, J. T. (1960). Interstage heat transfer and plate efficiency in a bubble plate column. Int. Symp. on Distillation, Inst. Chem. Engrs., Brighton, p. 171

English, G. E. and Van Winkle, M. (1963). Efficiency of fractionating columns, *Chem. Engng.* **70**, (23), 241

Fair, J. R. (1961). How to predict sieve tray entrainment and flooding, *Petro/Chem. Engineer* **33**, (10), 45

Fair, J. R. (1963). In Smith, B. D. *Design of Equilibrium Stage Processes*, Chap. 15. McGraw-Hill

Fair, J. R. (1970). Comparing trays and packings, *Chem. Engng. Prog.* **66**, (3), 45

Fair, J. R. (1983). Historical development of distillation equipment, *AIChE Symposium Series, Diamond Jubilee Historical/Review Volume*, p. 1

Fair, J. R. and Matthews, R. L. (1958). Better estimate of entrainment from bubble cap trays, *Pet. Ref.* **37**, 153

Fair, J. R., Null, H. R. and Bolles, W. L. (1983). Scale-up of plate efficiency from laboratory Oldershaw data, *Ind. Eng. Chem. Proc. Des. Dev.* **22**, 53

Fane, A. G. and Sawistowski, H. (1968). Surface tension effects in sieve-plate distillation, *Chem. Engng. Sci.* **23**, 943

Fane, A. G. and Sawistowski, H. (1969). Plate efficiencies in the foam and spray regimes of sieve-plate distillation, *Inst. Chem. Engrs. Symp. Series No. 32*, p. 1:8

Fane, A. G., Lindsey, J. K. and Sawistowski, H. (1973). Operation of a sieve plate in the spray regime of column operation, *Indian Chem. Engr.* **15**, (4), 37

Fell, C. J. D. and Pinczewski, W. V. (1977). New considerations in the design of high-capacity sieve trays, *The Chem. Engr.* (Jan.), 45

Fell, C. J. D. and Pinczewski, W. V. (1982). Coping with entrainment problems in low and moderate pressure distillation columns, *Inst. Chem. Engrs. Symp. Series No. 73*, p. D1

Finch, R. N. and Van Winkle, M. (1964). A statistical correlation of the efficiency of perforated trays, *Ind. Eng. Chem. Proc. Des. Dev.* **3**, (2), 107

Forbes, R. J. (1948). *A Short History of the Art of Distillation*. E. J. Brill, Leiden, Netherlands

Forgrieve, F. (1960). Commercial jet tray fractionators. Int. Symp. on Distillation, Inst. Chem. Engrs., Brighton, p. 185
Foss, A. S. (1957). PhD Thesis, University of Delaware
Francis, J. B. (1883). *Lowell Hydraulic Experiments*, 4th edn. Van Nostrand Company, New York
Frank, O. (1977). Shortcuts for distillation design, *Chem. Engng.* **84**, (6), 111
Friend, L., Lemieux, E. J. and Schreiner, W. C. (1960). New data on entrainment from perforated trays at close spacings, *Chem. Engng.* **67**, (22), 101
Gardner, R. G. and McLean, A. Y. (1969). Effect of system properties on sieve-plate froths, *Inst. Chem. Engrs. Symp. Series No. 32*, p. 2:39
Garner, F. H. and Porter, K. E. (1960). Mass transfer stages in distillation. Int. Symp. on Distillation, Inst. Chem. Engrs., Brighton, p. 43
Garvin, R. G. and Norton, E. R. (1968). Sieve tray performance under G.S. process conditions, *Chem. Eng. Prog.* **64**, (3), 99
Gautreaux, M. E. and O'Connell, H. E. (1955). Effect of length of liquid path on plate efficiency, *Chem. Engng. Prog.* **51**, (5), 232
Geddes, R. L. (1946). Local efficiencies of bubble plate fractionators, *Trans. Am. Inst. Chem. Engrs.* **42**, 79
Gerster, J. A., Hill, A. B., Hochgraf, N. N. and Robinson, D. G. (1958). Tray efficiencies in distillation columns. Final Report, University of Delaware, AIChE, New York
Gilbert, T. J. (1959). Liquid mixing on bubble cap and sieve plates, *Chem. Engng. Sci.* **10**, 243
Glitsch, Inc. (1974). *Ballast Tray Design Manual.* Glitsch, Inc., Dallas.
Goederen, C. W. J. de (1965). Distillation tray efficiency and interfacial area, *Chem. Engng. Sci.* **20**, 1115
Hai, N. T., Burgess, J. M., Pinczewski, W. V. and Fell, C. J. D. (1977). Mass transfer in the spray regime on an industrial sieve tray. 2nd Aust. Conf. on Heat & Mass Transfer, University of Sydney, Feb., p. 377
Harada, M., Adachi, M., Eguchi, W. and Nagata, S. (1964). Studies of fluid mixing on sieve plates, *Int. Chem. Engng.* **4**, (1), 165
Harris, I. J. (1965). Optimum design of sieve trays, *Brit. Chem. Engng.* **10**, (6), 377
Harris, I. J. and Roper, G. H. (1962). Performance characteristics of a 12 in. diameter sieve plate, *Can. J. Chem. Engng.* **40**, (6), 245
Hart, D. J. and Haselden, G. G. (1969). Influence of mixture composition on distillation-plate efficiency, *Inst. Chem. Engrs. Symp. Series No. 32*, p. 1:19
Haselden, G. G. (1975). Scope for improving fractionation equipment, *The Chem. Engnr.* (July/Aug.), 439
Haselden, G. G. and Thorogood, R. M. (1964). Point efficiency in the distillation of the oxygen–nitrogen–argon system, *Trans. Inst. Chem. Engrs.* **42**, T81
Haselden, G. G. and Witwit, S. (1981). Increasing column performance by the use of baffle trays, *Inst. Chem. Engrs. Symp Series No. 61*, p. 127
Haug, H. F. (1976). Stability of sieve trays with high overflow weirs, *Chem. Engng. Sci.* **31**, 295
Hausen, H. (1953). A definition of exchange efficiency of rectifying plates for binary and ternary mixtures, *Chem. Ing. Tech.* **25**, 595
Hay, J. M. and Johnson, A. I. (1960). A study of sieve tray efficiencies, *AIChE J.* **6**, (3), 373
Henderson, F. M. (1966). *Open Channel Flow.* Macmillan
Henley, E. J. and Seader, J. D. (1981). *Equilibrium-Stage Separation Operations in Chemical Engineering.* John Wiley & Sons
Higbie, R. (1935). The rate of absorption of a pure gas into a still liquid during short periods of exposure, *Trans. Am. Inst. Chem. Engrs.* **31**, 365
Hinze, J. O. (1965). Oscillations of a gas–liquid mixture on a sieve plate. Proc. Symp. on Two-Phase Flow, 21–23 June, University of Exeter, UK, p. F101

Ho, G. E., Muller, R. L. and Prince, R. G. H. (1969). Characterisation of two-phase flow patterns in plate columns, *Inst. Chem. Engrs. Symp. Series No. 32*, p. 2:10

Hobler, T. and Pawelczyk, P. (1972). Interfacial area in bubbling through a slot, *Brit. Chem. Engng. and Proc. Tech.* **17**, (7/8), 624

Hoek, P. J. and Zuiderweg, F. J. (1982). Influence of vapour entrainment on distillation tray efficiency at high pressures, *AIChE J.* **28**, (4), 535

Hoerner, B. K., Wiessner, F. G. and Berger, E. A. (1982). Effect of irregular motion of absorption/distillation processes, *Chem. Engng. Prog.* **78**, (11), 47

Hofer, H. (1983). Influence of gas-phase dispersion on plate column efficiency, *Ger. Chem. Eng.* **6**, 113

Hofhuis, P. A. M. (1980). Flow regimes on sieve trays for gas–liquid contacting. Doctoral Thesis, University of Technology, Delft

Hofhuis, P. A. M. and Zuiderweg, F. J. (1979). Sieve plates: dispersion density and flow regimes, *Inst. Chem. Engrs. Symp. Series No. 56*, p. 2.2/1

Holbrook, G. E. and Baker, E. M. (1934). Entrainment in a bubble cap distillation column, *Ind. Eng. Chem.* **26**, (10), 1063

Holland, C. D. (1963). *Multicomponent Distillation.* Prentice Hall

Holland, C. D. (1966). *Unsteady State Processes with Applications in Multicomponent Distillation.* Prentice Hall

Holland, C. D. (1975). *Fundamentals and Modeling of Separation Processes.* Prentice Hall

Holland, C. D. (1980). Computing large negative or positive values for the Murphree efficiencies, *Chem. Engng. Sci.* **35**, 2235

Holland, C. D. and McMahon, K. S. (1970). Comparison of vaporization efficiencies with Murphree-type efficiencies in distillation – I, *Chem. Engng. Sci.* **25**, 431

Honorat, A. and Sandall, O. C. (1978). Simultaneous heat and mass transfer in a packed binary distillation column, *Chem. Engng. Sci.* **33**, 635

Hovestreydt, J. (1963). The influence of the surface tension difference on the boiling of mixtures, *Chem. Engng. Sci.* **18**, 631

Huang, C. J. and Hodson, J. R. (1958). Perforated tray design procedure, *Pet. Ref.* **37**, (2), 104

Huesmann, K. (1966). Pressure loss and flow coefficients for flow vertical to perforated plates, *Chem. Ing. Tech.* **38**, (8), 877

Hughmark, G. A. (1971). Models for vapour phase and liquid phase mass transfer on distillation trays, *AIChE J.* **17**, (6), 1295

Hughmark, G. A. and O'Connell, H. E. (1957). Design of perforated plate fractionating towers, *Chem. Engng. Prog.* **53**, (3), 127

Hunt, C. d'A., Hanson, D. N. and Wilke, C. R. (1955). Capacity factors in the performance of perforated plate columns, *AIChE J.* **1**, (4), 441

Hutchinson, M. H. and Baddour, R. F. (1956). Ripple trays – a new tool for vapour–liquid contacting, *Chem. Engng. Prog.* **52**, (12), 503

Hutchinson, M. H., Buron, A. G. and Miller, B. P. (1949). Aerated flow principle applied to sieve plates. AIChE Meeting, 6–9 March, Los Angeles

Jackson, R. W. (1952). The formation of gas bubbles from orifices – Part 2, *The Industrial Chemist* **28**, (332), 391

Jamal, A. (1981). MSc Dissertation, University of Manchester Institute of Science and Technology

Jameson, G. J. and Kupferberg, A. (1967). Pressure behind a bubble accelerating from rest: simple theory and applications, *Chem. Engng. Sci.* **22**, 1053

Jenkins, A. E. O. (1981). The design of distillation columns for 'the times', *Inst. Chem. Engrs. Symp. Series No. 61*, p. 73

Jeromin, L., Holik, H. and Knapp, H. (1969). Efficiency calculation method for sieve-plate columns of air separation plants, *Inst. Chem. Engrs. Symp. Series No. 32*, p. 5:45

Jeronimo, M. A. and Sawistowski, H. (1973). Phase inversion correlation for sieve trays, *Trans. Inst. Chem. Engrs.* **51**, 265

Jeronimo, M. A. and Sawistowski, H. (1974). Projection velocities of droplets in the spray regime of sieve plate operation, *Trans. Inst. Chem. Engrs.* **52**, 291

Jeronimo, M. A. and Sawistowski, H. (1979). Characterisation of dispersion in the spray regime of sieve plate operation, *Inst. Chem. Engrs. Symp. Series No. 56*, p. 2.2/41

Jones, J. B. and Pyle, C. (1955). Relative performance of sieve and bubble cap plates, *Chem. Engng. Prog.* **51**, (9), 424

Kafarov, V. V., Shestopalov, V. V. and Bel'kov, V. P. (1972). Longitudinal mixing of liquid in a plate column with sieve trays for the absorption of nitric oxide, *Int. Chem. Engng.* **12**, (2), 257

Kafarov, V. V., Shestopalov, V. V. and Komissarov, Y. A. (1979). Vapour–liquid flow structure on bubbler plates, *Inst. Chem. Engrs. Symp. Series No. 56*, p. 2.3/79

Kaltenbacher, E. (1982). On the effect of the bubble size distribution and the gas-phase diffusion on the selectivity of sieve trays, *Chem. Engng. Fund.* **1**, (1), 47

Kamei, S. (1954). Minimum allowable vapour velocity for perforated plate columns, *Chem. Eng.* (*Tokyo*) **18**, 108

Kamei, S., Takamatsu, I. and Umeshita, I. (1951). Entrainment in perforated plate fractionating column, *Chem. Eng.* (*Japan*) **15**, 19

Kastanek, F. (1970). Efficiencies of different types of distillation plate, *Coll. Czech. Chem. Comm.* **35**, 1170

Kastanek, F. and Standart, G. (1966). *Efficiency of Distillation Trays.* Czech. Academy of Sciences Press, Prague

Kastanek, F. and Standart, G. (1967). Studies in distillation XX, *Sepn. Sci.* **2**, (4), 439

Katayama, H. and Imoto, T. (1972). Effect of vapour mixing on the tray efficiency of distillation columns, *J. Chem. Soc. Japan* (9), 1745

Kawagoe, K., Inoue, T., Nakao, K. and Otake, T. (1976). Flow pattern and gas holdup conditions in gas-sparged contactors, *Int. Chem. Engng.* **16**, (1), 176

Kayihan, F., Sandall, O. C. and Mellichamp, D. A. (1975). Simultaneous heat and mass transfer in binary distillation – I, *Chem. Engng. Sci.* **30**, 1333

Kayihan, F., Sandall, O. C. and Mellichamp, D. A. (1977). Simultaneous heat and mass transfer in binary distillation – II, *Chem. Engng. Sci.* **32**, 747

Kerner, H. T. (1976). *Foam Control Agents.* Noyes Data Corp., New Jersey

Khamdi, A. M., Skolbo, A. I. and Molokanov, Y. K. (1963). Some questions concerning the hydraulic gradients of plate columns, *Khim i Teknol Topliv i Masel* **8**, (2), 31

Kim, S. K. (1966). Theoretical study of vapour–liquid hold-up on a perforated plate, *Int. Chem. Engng.* **6**, (4), 634

King, C. J. (1980). *Separation Processes.* McGraw-Hill

Kirkpatrick, R. D. and Lockett, M. J. (1974). The influence of approach velocity on bubble coalescence, *Chem. Engng. Sci.* **29**, 2363

Kirschbaum, E. (1934). Efficiency of rectification and appropriate path for liquid flow, *Forsch. Gebiete Ingenieur* **5**, 245

Kirsten, R. A. and Van Winkle, M. (1970). Efficiency of jet trays, *Ind. Eng. Chem. Proc. Des. Dev.* **9**, (1), 100

Kister, H. Z. (1980). Column internals 1–7, *Chem. Engng.* **87**, (10), 138

Kister, H. Z., Pinczewski, W. V. and Fell, C. J. D. (1981*a*). The influence of operating parameters on entrainment from sieve trays. AIChE Meeting, April, Houston

Kister, H. Z., Pinczewski, W. V. and Fell, C. J. D. (1981*b*). Entrainment from sieve trays operating in the spray regime, *Ind. Eng. Chem. Proc. Des. Dev.* **20**, (3), 528

Kitchener, J. A. and Cooper, C. F. (1959). Current concepts in the theory of foaming, *Quart. Rev. of Chem. Soc.* (*London*) **13**, 71

Kneule, H. F. and Zelfel, E. (1966). Pressure drop of sieve plates when operating without liquid loss, *Chem. Ing. Tech.* **38**, (3), 260

Koch Engineering Co., Inc. (1982). *Flexitray Design Manual.* Koch Engineering Co., Inc., Wichita, Kansas

Koch, R. and Kuzniar, J. (1966). Hydraulic calculation of a weir sieve tray, *Int. Chem. Engng.* **6**, (4), 618

Kolar, V. (1969). The structure of gas–liquid mixtures on sieve trays of separation columns, *Chem. Engng. Sci.* **24**, 1285

Kolodzie, P. A. and Van Winkle, M. (1957). Discharge coefficients through perforated plates, *AIChE J.* **3**, (3), 305

Kouri, R. J. and Sohlo, J. J. (1985). Effect of developing liquid flow patterns on distillation plate efficiency, *Chem. Eng. Res. Des.* **63**, (Mar.), 117

Koziol, A. and Koch, R. (1973). Weeping of liquid and its effect on the operation of a bubble plate with downcomers, *Inzynieria Chemiczna* **3**, (4), 729

Koziol, A. and Koch, R. (1976). Hydraulics of large-hole sieve plates with downcomers, *Inzynieria Chemiczna* **6**, (3), 531

Kreis, H. and Raab, M. (1979). Industrial application of sieve trays with hole diameters from 1 mm to 25 mm with and without downcomers, *Inst. Chem. Engrs. Symp. Series No. 56*, p. 3.2/63

Krishna, R. (1976). Steady-state mass transport in multicomponent liquid mixtures, *Letters in Heat and Mass Transfer* **3**, (2), 153

Krishna, R. (1977). A film model analysis of non-equimolar distillation of multicomponent mixtures, *Chem. Engng. Sci.* **32**, 1197

Krishna, R. and Standart, G. L. (1979). Mass and energy transfer in multicomponent systems, *Chem. Eng. Commun.* **3**, 201

Krishna, R., Martinez, H. F., Shreedhar, R. and Standart, G. L. (1977). Murphree point efficiencies in multicomponent systems, *Trans. Inst. Chem. Engrs.* **55**, 178

Krummrich, K. (1984). Quoted in *Chemical Week* **134**, (24), 18

Kumar, R. and Kuloor, N. R. (1970). The formation of bubbles and drops, *Advances in Chem. Engng.* **8**, 256

Kuotsung, Y., Huang, J. and Fangzhen, G. (1983). Simulation and efficiency of a large tray. Selected papers of *J. Chem. Ind. & Eng. (China)* (2), June, 12

Kupferberg, A. and Jameson, G. J. (1969). Bubble formation at a submerged orifice above a gas chamber of finite volume, *Trans. Inst. Chem. Engrs.* **47**, T241

Kupferberg, A. and Jameson, G. J. (1970). Pressure fluctuations in a bubbling system with special reference to sieve plates, *Trans. Inst. Chem. Engrs.* **48**, T140

LaNauze, R. D. and Harris, I. J. (1972). On a model for the formation of gas bubbles at a single submerged orifice under constant pressure conditions, *Chem. Engng. Sci.* **27**, 2102

LaNauze, R. D. and Harris, I. J. (1974). Gas bubble formation at elevated system pressures, *Trans. Inst. Chem Engrs.* **52**, 337

Latimer, R. E. (1967). Distillation of air, *Chem. Engng. Prog.* **63**, (2), 35

Leibson, I., Holcomb, E. G., Cacoso, A. G. and Jacmic, J. J. (1956). Rate of flow and mechanics of bubble formation from single submerged orifices, *AIChE J.* **2**, (3), 296

Lemieux, E. J. (1983). Data for tower baffle design, *Hydrocarbon Processing* **62**, (9), 106

Lemieux, E. J. and Scotti, L. J. (1969). Perforated tray performance, *Chem. Engng. Prog.* **65**, (3), 52

Leva, M. (1972). Film trays for vacuum fractionation, *Chem. & Proc. Engng.* (July), 44

Lewis, W. K. (1936). Rectification of binary mixtures, *Ind. Engng. Chem.* **28**, (1), 399

Lim, C. T. (1973). PhD Thesis, University of Manchester Institute of Science and Technology

Lim, C. T., Porter, K. E. and Lockett, M. J. (1974). The effect of liquid channelling on two-pass distillation plate efficiency, *Trans. Inst. Chem. Engrs.* **52**, 193

Livansky, K. and Kolar, V. (1971). Hydrodynamics of plate columns, V, *Coll. Czech. Chem. Comm.* **36**, 286

Lockett, M. J. (1981). The froth to spray transition on sieve trays, *Trans. Inst. Chem. Engrs.* **59**, 26

Lockett, M. J. and Ahmed, I. S. (1983). Tray and point efficiencies from a 0.6 m diameter distillation column, *Chem. Eng. Res. Des.* **61**, (Mar.), 110

Lockett, M. J. and Banik, S. (1981). The effect of lateral liquid mixing in the downcomer on single pass distillation tray efficiency. *Proc. Chempor 81*, Povoa do Varzim, Portugal, p. 14

Lockett, M. J. and Banik, S. (1984). Weeping from sieve trays. AIChE Meeting, Nov., San Francisco

Lockett, M. J. and Dhulesia, H. A. (1980). Murphree plate efficiency with non-uniform vapour distribution, *Chem. Engng. J.* **19**, 183

Lockett, M. J. and Gharani, A. A. W. (1979). Downcomer hydraulics at high liquid flow rates, *Inst. Chem. Engrs. Symp. Series No. 56*, p. 2.3/43

Lockett, M. J. and Kirkpatrick, R. D. (1975). Ideal bubbly flow and actual flow in bubble columns, *Trans. Inst. Chem. Engrs.* **53**, 267

Lockett, M. J. and Plaka, T. (1983). Effect of non-uniform bubbles in the froth on the correlation and prediction of point efficiencies, *Chem. Eng. Res. Des.* **61**, (Mar.), 119

Lockett, M. J. and Safekourdi, A. (1976). The effect of the liquid flow pattern on distillation plate efficiency, *Chem. Engng. J.* **11**, 117

Lockett, M. J. and Safekourdi, A. (1977). Light transmission through bubble swarms, *AIChE J.* **23**, (3), 395

Lockett, M. J. and Uddin, M. S. (1978). Slotted distillation trays: momentum transfer from a single slot, *Trans. Inst. Chem. Engrs.* **56**, 194

Lockett, M. J., Kirkpatrick, R. D. and Uddin, M. S. (1979). Froth regime point efficiency for gas-film controlled mass transfer on a two-dimensional sieve tray, *Trans. Inst. Chem. Engrs.* **57**, 25

Lockett, M. J., Lim, C. T. and Porter, K. E. (1973). The effect of liquid channelling on distillation column efficiency in the absence of vapour mixing, *Trans. Inst. Chem. Engrs.* **51**, 61

Lockett, M. J., Plaka, T. and Ahmed, I. S. (1984). Scaling up distillation tray efficiency under conditions of weeping or entrainment when liquid channelling also occurs, *Chem. Eng. Res. Des.* **62**, (3), 191

Lockett, M. J., Porter, K. E. and Bassoon, K. S. (1975). The effect of vapour mixing on distillation plate efficiency when liquid channelling occurs, *Trans. Inst. Chem. Engrs.* **53**, 125

Lockett, M. J., Rahman, M. A. and Dhulesia, H. A. (1983). The effect of entrainment on distillation tray efficiency, *Chem. Engng. Sci.* **38**, 661

Lockett, M. J., Rahman, M. A. and Dhulesia, H. A. (1984). Prediction of the effect of weeping on distillation tray efficiency, *AIChE J.* **30**, (3), 423

Lockett, M. J., Spiller, G. T. and Porter, K. E. (1976). The effect of the operating regime on entrainment from sieve trays, *Trans. Inst. Chem. Engrs.* **54**, 202

Lockhart, F. J. and Leggett, C. W. (1958). New fractionating tray designs. In *Advances in Petroleum Refining and Chemistry*, eds. K. A. Kobe and J. J. McKetta, vol. 1, p. 277. Interscience, New York

Loon, R. E., Pinczewski, W. V. and Fell, C. J. D. (1973). Dependence of the froth-to-spray transition on sieve tray design parameters, *Trans. Inst. Chem. Engrs.* **51**, 374

Lowry, R. P. and Van Winkle, M. (1969). Foaming and frothing related to system physical properties in a small perforated plate distillation column, *AIChE J.* **15**, (5), 665

McAllister, R. A., McGinnis, P. H. and Plank, C. A. (1958). Perforated plate performance, *Chem. Engng. Sci.* **9**, 25; (1961). **13**, 269

McCann, D. J. and Prince, R. G. H. (1969). Bubble formation and weeping at a submerged orifice, *Chem. Engng. Sci.* **24**, 801

McCann, D. J. and Prince, R. G. H. (1971). Regimes of bubbling at a submerged orifice, *Chem. Engng. Sci.* **26**, 1505

MacFarland, S. A., Sigmund, P. M. and Van Winkle, M. (1972). Predict distillation efficiency, *Hydrocarbon Processing* **51**, (July), 111

McLachlan, C. N. S. and Danckwerts, P. V. (1972). Desorption of carbon dioxide from aqueous potash solutions with and without the addition of arsenite as a catalyst, *Trans. Inst. Chem. Engrs.* **50**, 300
McNeil, K. M. (1970). The measurement of interfacial area and surface-renewal rate in a bubble cap plate, *Can. J. Chem. Engng.* **48**, (June), 252
Malafeev, N. A. and Malyusov, V. A. (1973). Effect of liquid entrainment on the efficiency of tubular columns with cocurrent phase interaction, *Theo. Found. Chem. Engng.* **7**, 101
Manickampillai, M. and Sawistowski, H. (1981). Relationship between plate and point efficiencies for plates operating in the spray regime. *Proc. Chempor 81*, Povoa do Varzim, Portugal, p. 1.65
Manning, E. (1964). High capacity distillation trays, *Ind. Eng. Chem.* **56**, (4), 14
Marangoni, C. (1871). Spreading of droplets of a liquid on the surface of another, *Ann. Phys. Lpz.* **143**, (7), 337
Mayfield, F. D., Church, W. L., Green, A. C., Lee, D. C. and Rasmussen, R. W. (1952). Perforated plate distillation columns, *Ind. Engng. Chem.* **44**, (9), 2238
Mecklenburgh, J. C. (1974). Backmixing and design: a review, *Trans. Inst. Chem. Engrs.* **52**, 180
Medina, A. G., Ashton, N. and McDermott, C. (1978). Murphree and vaporization efficiencies in multicomponent distillation, *Chem. Engng. Sci.* **33**, 331
Medina, A. G., Ashton, N. and McDermott, C. (1979*a*). Hausen and Murphree efficiencies in binary and multicomponent distillation, *Chem. Engng. Sci.* **34**, 1105
Medina, A. G., McDermott, C. and Ashton, N. (1979*b*). Prediction of multicomponent distillation efficiencies, *Chem. Engng. Sci.* **34**, 861
Mehta, V. D. and Sharma, M. M. (1966). Effect of diffusivity on gas-side mass transfer coefficient, *Chem. Engng. Sci.* **21**, 361
Mersmann, A. (1963). When are all the holes in a sieve plate employed during flow?, *Chem. Ing. Tech.* **35**, (2), 103
Mikhailenko, G. G., Bol'shakov, A. G., Ennan, G. G. and Gansh, A. I. (1975). Spray entrainment from grid plates under increased loads, *Vopr. Khimii. I. Khim Teknol. Resp. Mezhved. Termat. Nauch-Tekhn. Sb.* **37**, 77
Miyahara, T., Haga, N. and Takahashi, T. (1983*a*). Bubble formation from an orifice at high flow rates, *Int. Chem. Engng.* **23**, (3), 524
Miyahara, T., Matsuba, Y. and Takahashi, T. (1983*b*). The size of bubbles generated from perforated plates, *Int. Chem. Engng.* **23**, (3), 517
Moens, F. P. and Bos, R. G. (1972). Surface renewal effects in distillation, *Chem. Engng. Sci.* **27**, 403
Molnar, K. (1974). Eddy-diffusion coefficient in valve tray distillation columns, *Periodica Polytechnica Mechanical Engineering* **18**, (2–3), 155
Molokanov, Y. K. (1963). The hydraulic resistance of grid and perforated downcomerless trays, *Int. Chem. Eng.* **3**, (2), 157
Molokanov, Y. K., Korablina, T. P., Abushevich, I. Z. and Rogozina, L. P. (1969). The effect of irregularity of liquid entrainment by the flow of gas over a tray on the results of measurement in the distribution of entrainment by different methods, *Int. Chem. Engng.* **9**, (4), 603
Muhle, J. (1972). Calculation of the dry pressure drop across perforated plates, *Chem. Ing. Tech.* **44**, (1/2), 72
Mukhlenov, I. P. and Tarat, E. Y. (1958). Hydraulic resistance of sieve plates, *Zh. Prikl. Khim.* **31**, 542
Muller, R. L. and Prince, R. G. H. (1972). Regimes of bubbling and jetting from submerged orifices, *Chem. Engng. Sci.* **27**, 1583
Murphree, E. V. (1925). Rectifying column calculations with particular reference to *N* component mixtures, *Ind. Engng. Chem.* **17**, (7), 747

Neretnieks, I. (1970). The optimisation of sieve tray columns, *Brit. Chem. Engng.* **15**, (2), 193

Neuburg, H. J. and Chuang, K. T. (1982). Mass transfer modelling for GS heavy water plants, *Can. J. Chem. Engng.* **60**, (Aug.), 504 & 510

Newitt, D. M., Dombrowski, N. and Knelman, F. H. (1954). Liquid entrainment 1. The mechanism of drop formation from gas or vapour bubbles, *Trans. Inst. Chem. Engrs.* **32**, 244

Nielsen, R. D., Tek, M. R. and York, J. L. (1965). Mechanism of entrainment formation in distillation columns. Proc. Symposium on Two Phase Flow, 21–23 June, University of Exeter, UK, p. F201

Nord, M. (1946). Temperature efficiency in distillation, *Ind. Engng. Chem.* **38**, (6), 657

Norman, W. S. (1960). Int. Symp. on Distillation. Inst. Chem. Engrs., Brighton. Contribution to discussion, p. 33

Nutter Engineering Co. (1976). *Float Valve Design Manual.* Nutter Engineering Co., Tulsa, Oklahoma

Nutter, D. E. (1971). Ammonia stripping efficiency studies. Paper 49c, AIChE 68th National Meeting, Houston, Texas

Nutter, D. E. (1972). Ammonia stripping efficiency studies, *AIChE Symp. Series No. 124*, **68**, 73

Nutter, D. E. (1979). Weeping and entrainment studies for sieve and V-grid trays in an air–oil system, *Inst. Chem. Engrs. Symp. Series No. 56*, p. 3.2/47

O'Connell, H. E. (1946). Plate efficiency of fractionating columns and absorbers, *Trans. Am. Inst. Chem. Engrs.* **42**, 741

Pasiuk-Bronikowska, W. (1969). Attempts to determine the liquid-film coefficient for physical absorption and effective interfacial area in a sieve plate column by the chemical method, *Chem. Engng. Sci.* **24**, 1139

Payne, G. J. and Prince, R. G. H. (1975). The transition from jetting to bubbling at a submerged orifice, *Trans. Inst. Chem. Engrs.* **53**, 209

Payne, G. J. and Prince, R. G. H. (1977). The relationship between the froth and spray regimes, and the orifice processes occurring on perforated distillation plates, *Trans. Inst. Chem. Engrs.* **55**, 266

Perry, R. H. and Green, D. W. (1984). *Perry's Chemical Engineers' Handbook*, 6th edn. McGraw-Hill

Pinczewski, W. V. (1981). The formation and growth of bubbles at a submerged orifice, *Chem. Engng. Sci.* **36**, 405

Pinczewski, W. V. and Fell, C. J. D. (1971). Droplet projection velocities for use in sieve tray spray models, *Can. J. Chem. Engng.* **49**, (Aug.), 548

Pinczewski, W. V. and Fell, C. J. D. (1972). The transition from froth-to-spray regime on commercially loaded sieve trays, *Trans. Inst. Chem. Engrs.* **50**, 102

Pinczewski, W. V. and Fell, C. J. D. (1974). Nature of the two-phase dispersion on sieve plates operating in the spray regime, *Trans. Inst. Chem. Engrs.* **52**, 294

Pinczewski, W. V. and Fell, C. J. D. (1975). Oscillations on sieve trays, *AIChE J.* **21**, (5), 1019

Pinczewski, W. V. and Fell, C. J. D. (1977). Droplet sizes on sieve plates operating in the spray regime, *Trans. Inst. Chem. Engrs.* **55**, 46

Pinczewski, W. V., Benke, N. D. and Fell, C. J. D. (1975). Phase inversion on sieve trays, *AIChE J.* **21**, (6), 1210

Pinczewski, W. V., Yeo, H. K. and Fell, C. J. D. (1973). Transition behaviour at submerged orifices, *Chem. Engng. Sci.* **28**, 2261

Piqueur, H. and Verhoeye, L. (1976). Research on valve trays. Hydraulic performance in the air–water system, *Can. J. Chem. Engng.* **54**, (June), 177

Pohjola, V. J. (1973). Point efficiency in binary distillation with unequal molar fluxes of components, *Ind. Eng. Chem. Proc. Des. Dev.* **12**, (3), 398

Pohorecki, R. (1968). The absorption of CO_2 in carbonate–bicarbonate buffer solutions containing hypochlorite catalyst on a sieve plate, *Chem. Engng. Sci.* **23**, 1447

Pohorecki, R. (1976). Mass transfer with chemical reaction during gas absorption on a sieve plate, *Chem. Engng. Sci.* **31**, 637

Porter, K. E. and Jenkins, J. D. (1979). The interrelationship between industrial practice and academic research in distillation and absorption, *Inst. Chem. Engrs. Symp. Series No. 56*, Discussion volume, p. 75

Porter, K. E. and Wong, P. F. Y. (1969). Transition from spray to bubbling on sieve plates, *Inst. Chem. Engrs. Symp. Series No. 32*, p. 2:22

Porter, K. E., Davies, B. T. and Wong, P. F. Y. (1967). Mass transfer and bubble sizes in cellular foams and froths, *Trans. Inst. Chem. Engrs.* **45**, T265

Porter, K. E., King, M. B. and Varshney, K. C. (1966). Interfacial areas and liquid–film mass transfer coefficients for a 3 ft. diameter bubble-cap plate, *Trans. Inst. Chem. Engrs.* **44**, T274

Porter, K. E., Lockett, M. J. and Lim, C. T. (1972). The effect of liquid channelling on distillation plate efficiency, *Trans. Inst. Chem. Engrs.* **50**, 91

Porter, K. E., O'Donnell, K. A. and Latifipour, M. (1982). The use of water cooling to investigate flow pattern effects on tray efficiency, *Inst. Chem. Engrs. Symp. Series No. 73*, p. L33

Porter, K. E., Safekourdi, A. and Lockett, M. J. (1977). Plate efficiency in the spray regime, *Trans. Inst. Chem. Engrs.* **55**, 190

Potter, O. E. (1969). Bubble formation under constant pressure conditions, *Chem. Engng. Sci.* **24**, 1733

Pozin, M. E., Mukhlenov, I. P. and Tarat, E. Y. (1957). Nature of the gas–liquid disperse system, *J. Appl. Chem. USSR* **30**, (1), 43

Pratt, C. F. and Hobbs, S. Y. (1975). Quick kill of foams on fractionator trays, *Chem. Engng.* **82**, (10), 112

Priestman, G. H. and Brown, D. J. (1981). The mechanism of pressure pulsation in sieve tray columns, *Trans. Inst. Chem. Engrs.* **59**, 279

Prince, R. G. H. (1960). Characteristics and design of perforated plates. Int. Symp. on Distillation, Inst. Chem. Engrs., Brighton, p. 177

Prince, R. G. H. and Chan, B. K. C. (1965). The seal point of perforated distillation plates, *Trans. Inst. Chem. Engrs.* **43**, T49

Prince, R. G. H., Jones, A. P. and Panic, R. J. (1979). The froth spray transition, *Inst. Chem. Engrs. Symp. Series No. 56*, p. 2.2/27

Ramakrishnan, S., Kumar, R. and Kuloor, N. R. (1969). Studies in bubble formation – I. Bubble formation under constant flow conditions, *Chem. Engng. Sci.* **24**, 731

Ramm, V. M. (1968). *Absorption of Gases.* Israel Program for Scientific Translation Press, Jerusalem

Raper, J. A., Burgess J. M. and Fell, C. J. D. (1977*a*). Frothy dispersion characteristics on industrial scale sieve trays. Inst. Chem. Engrs. 4th Annual Research Meeting, Swansea

Raper, J. A., Phuong, T. V. and Fell, C. J. D. (1977*b*). Strategy for the design of optimal industrial sieve trays. Chemeca 77, 14–16 Sept., Canberra

Raper, J. A., Pinczewski, W. V. and Fell, C. J. D. (1984). Liquid passage on sieve trays operating in the spray regime, *Chem. Eng. Res. Des.* **62**, (Mar.), 111

Raper, J. A., Dixon, D. C., Fell, C. J. D. and Burgess, J. M. (1978). Limitations of Burgess–Calderbank probe for characterisation of gas–liquid dispersions on sieve trays, *Chem. Engng. Sci.* **33**, 1405

Raper, J. A., Hai, N. T., Pinczewski, W. V. and Fell, C. J. D. (1979). Mass transfer efficiency on simulated industrial sieve trays operating in the spray regime, *Inst. Chem. Engrs. Symp. Series No. 56*, p. 2.2/57

Raper, J. A., Kearney, M. S., Burgess, J. M. and Fell, C. J. D. (1982). The structure of industrial sieve tray froths, *Chem. Engng. Sci.* **37**, 501

Raskop, F. (1974). Perform Kontakt plates, *The Chem. Eng.* (Nov.), 709

Reid, R. C., Prausnitz, J. M. and Sherwood, T. K. (1977). *The Properties of Gases and Liquids*, 3rd edn. McGraw-Hill

Rennie, J. and Evans, F. (1962). The formation of froths and foams above sieve plates, *Brit. Chem. Engng.* **7**, (7), 498

Rennie, J. and Smith, W. (1965). A photographic study of the formation and properties of large gas bubbles and their breakdown into froths. Inst. Chem. Engrs./AIChE Transport Phenomena Symposium, p. 62

Rodionov, A. I. and Radikovskii, V. M. (1967). Distribution of gas bubbles in the two-phase layer on bubbler plates, *Zhu. Prikl. Khimii* **40**, (12), 2751

Rodionov, A. I. and Vinter, A. A. (1967). Investigation of interfacial surface area on sieve trays by a chemical method, *Int. Chem. Engng.* **7**, (3), 468

Ross, S. (1967). Mechanisms of foam stabilization and antifoaming action, *Chem. Engng. Prog.* **63**, (9), 41

Ross, S. and Nishioka, G. (1975). Foaming behaviour of partially miscible liquids as related to their phase diagrams. Int. Conf. on Foams, 8–10 Sept. Society of Chemical Industry (London), Brunel University

Ruckenstein, E. (1970). The interaction between heat and mass transfer in the rectification of mixtures, *AIChE J.* **16**, (1), 144

Ruckenstein, E. and Smigelschi, O. (1965). On thermal effects in rectification of mixtures, *Chem. Engng. Sci.* **20**, 66

Ruckenstein, E. and Smigelschi, O. (1967). The thermal theory and the plate efficiency, *Can. J. Chem. Engng.* **45**, (Dec.), 334

Ruff, K., Pilhofer, T. and Mersmann, A. (1978). Ensuring flow through all the openings of perforated plates for fluid dispersion, *Int. Chem. Engng.* **18**, (3), 395

Rush, F. E. and Stirba, C. (1957). Measured plate efficiencies and values predicted from single-phase studies, *AIChE J.* **3** (3), 336

Rylek, M. and Standart, G. (1964). The hydraulics of sieve trays, *Int. Chem. Engng.* **4**, (4), 711

Sagert, N. H. and Quinn, M. J. (1976). The coalescence of H_2S and CO_2 bubbles in water, *Can. J. Chem. Engng.* **54**, (Oct.), 392

Sakata, M. and Yanagi, T. (1979). Performance of a commercial sieve tray, *Inst. Chem. Engrs. Symp. Series No. 56*, p. 3.2/21

Sandall, O. C. and Dribika, M. M. (1979). Simultaneous heat and mass transfer for multicomponent distillation in continuous contact equipment, *Inst. Chem. Engrs. Symp. Series No. 56*, p. 2.5/1

Sargent, R. W. H. and MacMillan, W. P. (1962). The physical properties of foams and froths, *Trans. Inst. Chem. Engrs.* **40**, 191

Sargent, R. W. H. and Murtagh, B. A. (1969). The design of plate distillation columns for multicomponent mixtures, *Trans. Inst. Chem. Engrs.* **47**, T85

Sargent, R. W. H., Bernard, J. D. T., MacMillan, W. P. and Schroter, R. C. (1964). The performance of sieve-plates in distillation. In *Distillation – Final Report ABCM/BCPMA Distillation Panel*, p. 81. Chemical Industries Association, London

Satyanarayan, A., Kumar, R. and Kuloor, N. R. (1969). Studies in bubble formation – II. Bubble formation under constant pressure conditions, *Chem. Engng. Sci.* **24**, 749

Sawistowski, H. (1978). Hydrodynamics and mass transfer behaviour of sieve plates in the spray and bubble regimes, *Chem. Ing. Tech.* **50**, 743

Sawistowski, H., Bainbridge, G. S., Stacey, M. J. and Theobald, A. (1964). Some aspects of heat and mass transfer in distillation. In *Distillation – Final Report ABCM/BCPMA Distillation Panel*, p. 143. Chemical Industries Association, London

Scali, C. and Zanelli, S. (1982). A characterisation of oscillating phenomena on sieve plates of transfer columns, *Chem. Engng. J.* **25**, 191

Sealey, C. J. (1970). The optimal design of a laboratory distillation column for efficiency studies at finite reflux, *Chem. Engng. Sci.* **25**, 561
Sewell, A. (1975). Practical aspects of distillation column design, *The Chem. Engr.* (July/Aug.), 442
Shakov, Y. A., Noskov, A. A. and Romankov, P. G. (1964). The upper limit of foaming on sieve plates, *Zhu. Prikl. Khimii* **37**, (9), 2074
Sharma, M. M. and Danckwerts, P. V. (1970). Chemical methods of measuring interfacial area and mass transfer coefficients in two-fluid systems. *Brit. Chem. Engng.* **15**, (4), 522
Sharma, M. M. and Gupta, R. K. (1967). Mass transfer characteristics of plate columns without downcomers, *Trans. Inst. Chem. Engrs.* **45**, T169
Sharma, M. M., Mashelkar, R. A. and Mehta, V. D. (1969). Mass transfer in plate columns, *Brit. Chem. Engng.* **14**, (1), 70
Shore, D. and Haselden, G. G. (1969). Liquid mixing on distillation plates and its effect on plate efficiency, *Inst. Chem. Engrs. Symp. Series No. 32*, p. 2:54
Silvey, F. C. and Keller, G. J. (1966). Testing on a commercial scale, *Chem. Engng. Prog.* **62**, (1), 68
Silvey, F. C. and Keller, G. J. (1969). Performance of 3 sizes of ceramic Raschig rings in a 4 ft. diameter column, *Inst. Chem. Engrs. Symp. Series No. 32*, p. 4:18
Smith, B. D. (1963). *Design of Equilibrium Stage Processes.* McGraw-Hill
Smith, V. C. and Delnicki, W. V. (1975). Optimum sieve tray design, *Chem. Engng. Prog.* **71**, (8), 68
Smith, P. L. and Van Winkle, M. (1958). Discharge coefficients through perforated plates at Re of 400 to 3000, *AIChE J.* **4**, (3), 266
Smith, R. K. and Wills, G. B. (1966). Application of penetration theory to gas absorption on a sieve tray, *Ind. Eng. Chem. Proc. Des. Dev.* **5**, (1), 39
Sohlo, J. J. and Kinnunen, S. (1977). Dispersion and flow phenomena on a sieve plate, *Trans. Inst. Chem. Engrs.* **55**, 71
Sohlo, J. J. and Kouri, R. J. (1982). An analysis of enhanced transverse dispersion on distillation plates, *Chem. Engng. Sci.* **37**, 193
Solari, R. B. and Bell, R. L. (1978). The effect of transverse eddy dispersion on distillation tray efficiency. AIChE Meeting, Feb., Atlanta
Solari, R. B., Saez, E., D'apollo, I. and Bellet, A. (1982). Velocity distribution and liquid flow patterns on industrial sieve trays, *Chem. Eng. Commun.* **13**, 369
Spells, K. E. (1954). Bubble formation by rapid air flow through slots submerged in water, *Trans. Inst. Chem. Engrs.* **32**, 167
Spells, K. E. and Bakowski, S. (1950). A study of bubble formation at single slots submerged in water, *Trans. Inst. Chem. Engrs.* **28**, 38
Spells, K. E. and Bakowski, S. (1952). The formation of bubbles at closely spaced slots submerged in water, *Trans. Inst. Chem. Engrs.* **30**, 189
Standart, G. (1965). Studies on distillation – V, *Chem. Engng. Sci.* **20**, 611
Standart, G. (1971). Comparison of Murphree-type efficiencies with vaporisation efficiencies, *Chem. Engng. Sci.* **26**, 985
Standart, G., Bragg, R., Uddin, M. S., El Yafi, A. H. and Yaroson, E. (1979). New models for point efficiency, *Inst. Chem. Engrs. Symp. Series No. 56*, p. 2.1/1
Stanislas, D. H. and Smith, A. J. (1960). Performance of distillation contacting devices, Int. Symp. on Distillation, Inst. Chem. Engrs., Brighton, p. 208
Steiner, L., Ballmer, J. F. and Hartland, S. (1975). The structure of gas–liquid dispersions on perforated plates, *Chem. Engng. J.* **10**, 35
Steiner, L., Hunkeler, R. and Hartland, S. (1977). Behaviour of dynamic cellular foams, *Trans. Inst. Chem. Engrs.* **55**, 153
Sterbacek, Z. (1964). Pressure loss of perforated plates, *Chem. Prumysl* **14**, (3), 119
Sterbacek, Z. (1967). Hydrodynamics of perforated trays with downcomers, *Brit. Chem. Engng.* **12**, (10), 1577

Sterbacek, Z. (1968). Liquid dispersion coefficients on perforated plates with downcomers, *Trans. Inst. Chem. Engrs.* **46**, T167

Stichlmair, J. (1978). *Bodenkolonne.* Verlag Chemie

Stichlmair, J. and Mersmann, A. (1978). Dimensioning plate columns for absorption and rectification, *Int. Chem. Engng.* **18**, (2), 223

Stichlmair, J. and Weisshuhn, E. (1973). Studies of plate efficiency paying particular attention to liquid mixing, *Chem. Ing. Tech.* **45**, (5), 242

Strand, C. P. (1963). Bubble cap tray efficiencies, *Chem. Engng. Prog.* **59**, (4), 58

Strang, L. C. (1934). Entrainment in a bubble-cap fractionating column, *Trans. Inst. Chem. Engrs.* **12**, 169

Sum-Shik, L. E., Aerov, M. E. and Bystrova, J. A. (1963). Entrainment in, and hydrodynamic design of, columns with downcomerless trays, *Khim. Prom.* **1**, 63

Tadaki, S. and Maeda, S. (1963). The size of bubbles from perforated plates, *Chem. Eng. (Japan)* **1**, 106

Takahashi, T., Matsuno, R. and Miyahara, T. (1973). Theoretical consideration of gas void fraction and height on a perforated plate, *J. Chem. Engng. Japan* **6**, (1), 38

Takahashi, T., Miyahara, T. and Sato, T. (1979). Gas void fraction on a perforated plate, *J. Chem. Engng. Japan* **12**, (4), 269

Takeuchi, H., Takahashi, K. and Kizawa, N. (1977). Absorption of nitrogen dioxide in sodium sulphate solution from air as a diluent, *Ind. Eng. Chem. Proc. Des. Dev.* **16**, (4), 486

Tasev, Z. and Elenkov, D. (1970). Comparative study of entrainment on valve plates, *God., Vissh., Khim-Tekhnol, Inst. Burgas. Bulg.* **7**, (7), 121

Teller, A. J., Cheng, S. I. and Davies, H. A. (1963). Protruded sieve tray performance, *AIChE J.* **9**, (3), 407

Thibodeaux, L. J. and Murrill, P. W. (1966). Comparing packed and plate columns, *Chemical Engineering* **73**, (15), 155

Thomas, W. J. and Campbell, M. (1967). Hydraulic studies in a sieve plate downcomer system, *Trans. Inst. Chem. Engrs.* **45**, T53

Thomas, W. J. and Haq, M. A. (1976). Studies of the performance of a sieve tray with 3/8 in. diameter perforations, *Ind. Eng. Chem. Proc. Des. Dev.* **15**, (4), 509

Thomas, W. J. and Ogboja, O. (1978). Hydraulic studies in sieve tray columns, *Ind. Eng. Chem. Proc. Des. Dev.* **17**, (4), 429

Thomas, W. J. and Shah, A. N. (1964). Downcomer studies in a frothing system, *Trans. Inst. Chem. Engrs.* **42**, T71

Todd, W. G. and Van Winkle, M. (1972). Correlation of valve tray efficiency data, *Ind. Eng. Chem. Proc. Des. Dev.* **11**, (4), 589

Toor, H. L. and Burchard, J. K. (1960). Plate efficiencies in multicomponent distillation, *AIChE J.* **6**, (2), 202

Treybal, R. E. (1968). *Mass Transfer Operations*, 2nd edn. McGraw-Hill

Union Carbide Corp. (1970). *MD-Tray Design Manual.* Union Carbide Corp., Tonawanda, New York

Unno, H. and Inoue, I. (1976). Froth density profiles and froth height on perforated plate, *J. Chem. Engng. Japan* **9**, (2), 92

Val'dberg, A. Y., YaTarat, E. and Zaitsev, M. M. (1969). Investigation of liquid entrainment from foam apparatus with grid plates, *Int. Chem. Engng.* **9**, (4), 638

Van Winkle, M. (1967). *Distillation.* McGraw-Hill

Vikhman, A. G., Berkovskii, M. A. and Kruglov, S. A. (1976). Study of the effect of the width of overflow on the operation of mass transfer plate under cross current conditions, *Khim. Neft. Mash.* **3**, 17

Vital, T. J., Grossel, S. S. and Olsen, P. I. (1984). Estimating separation efficiency – plate columns. *Hydrocarbon Processing* **63**, (11), 147

Vogelpohl, A. (1979). Murphree efficiencies in multicomponent systems, *Inst. Chem. Engrs. Symp. Series No. 56*, p. 2.1/25

Vybornov, V. G., Aleksandrov, I. A. and Zykov, D. D. (1971). The effect of transverse maldistribution of vapour and liquid flows on the operating efficiency of crossflow trays, *Teor. Osnovy. Khim. Tekh.* **5**, (6), 779

van der Meer, D. (1971). Foam stabilisation in small and large bubble columns. Joint Meeting on Bubbles and Foams, Verfahrenstechnische Gesellschaft im VDI/Inst. Chem. Engrs., Sept., Nurenberg, p. S2-4.1

van der Meer, D., Zuiderweg, F. J. and Scheffer, H. J. (1971). Foam suppression in extract purification and recovery trains. Poc. Int. Solvent Extraction Conf., The Hague, Society of Chemical Industry (London), paper 124, p. 28

Wada, T., Kageyama, O. and Azami, S. (1966). Characteristics of gas–liquid contact and tray efficiency on perforated tray, *Chem. Eng.* (*Tokyo*) **30**, 507

Wallis, G. B. (1969). *One-Dimensional Two-Phase Flow*. McGraw-Hill

Wallis, G. B. and Kuo, J. T. (1976). The behaviour of gas–liquid interfaces in vertical tubes, *Int. J. Multiphase Flow* **2**, 521

Walter, J. F. and Sherwood, T. K. (1941). Gas absorption in bubble-cap columns, *Ind. Eng. Chem.* **33**, (4), 493

Wehner, J. F. and Wilhelm, R. H. (1956). Boundary conditions of flow reactor, *Chem. Engng. Sci.* **6**, 89

Weiler, D. W., Kirkpatrick, R. D. and Lockett, M. J. (1981). Effect of downcomer mixing on distillation tray efficiency, *Chem. Engng. Prog.* **77**, (1), 63

Weiler, D. W., Delnicki, W. V. and England, B. L. (1973). Flow hydraulics of large diameter trays, *Chem. Eng. Prog.* **69**, (10), 67

Weiss, S. and Langer, J. (1979). Mass transfer on valve trays with modifications of the structure of dispersions, *Inst. Chem. Engrs. Symp. Series No. 56*, p. 2.3/1

Welch, N. E., Durbin, L. D. and Holland, C. D. (1964). Mixing on valve trays and in downcomers of a distillation column, *AIChE J.* **10**, (3), 373

West, F. B., Gilbert, W. D. and Shimizu, T. (1952). Mechanism of mass transfer on bubble plates, *Ind. & Engng. Chem.* **44**, (10), 2470

Wilke, C. R. (1950). Diffusional properties of multicomponent gases, *Chem. Engng. Prog.* **46**, 95

Williams, B., Begley, J. W. and Wu, C. (1960). Tray efficiencies in distillation columns. Final report from the University of Michigan, AIChE, New York

Winter, G. R. and Uitti, K. D. (1976). Froth initiators can improve tray performance, *Chem. Engng. Prog.* **72**, (9), 50

Wong, P. F. Y. (1967). PhD Thesis, University of Birmingham

Wong, P. F. Y. and Kwan, W. K. (1979). A generalised method for predicting the spray-bubbling transition on sieve plates, *Trans. Inst. Chem. Engrs.* **57**, 205

Wood, R. M. (1961). The stability of the valves of floating valve plates, *Trans. Inst. Chem. Engrs.* **39**, 313

Wraith, A. E. (1971). Two stage bubble growth at a submerged plate orifice, *Chem. Engng. Sci.* **26**, 1659

Yanagi, T. and Scott, B. D. (1973). Effects of liquid mixing on commercial scale sieve tray efficiency. AIChE Meeting, March, New Orleans

Ying, D. H. S., Thorogood, R. M. and Fox, V. G. (1984). Performance of circular flow and crossflow distillation trays with maldistribution and weeping. AIChE Meeting, Nov., San Francisco

Zanelli, S. (1975). Perforated plate weeping: The influence of different types of weir, *Quad. Ing. Chim. Ital.* **11**, (3), 41

Zelfel, E. (1965). Study of resistance coefficient of a single orifice, *Chem. Ing. Tech.* **37**, (12), 1209

Zelfel, E. (1967). The pressure drop across the liquid layer of a sieve plate, *Chem. Ing. Tech.* **39**, (5/6), 284
Zenz, F. A., Stone, L. and Crane, M. (1967). Find sieve tray weepage rates, *Hydrocarbon Processing* **46**, (12), 138
Zhavoronkov, N. M., Malyusov, V. A. and Zel'vensky, Ya. D. (1979). Prediction of rectification kinetics incorporating heat exchange between phases, *Inst. Chem. Engrs. Symp. Series No. 56*, p. 2.1/33
Zhou, Y. F., Shi, J. F., Wang, X. M. and Ye, Y. H. (1980). A study of the hydrodynamic behaviour of Linde-type flow-guided sieve plates, *Int. Chem. Engng.* **20**, (4), 642
Zuiderweg, F. J. (1973). Distillation – science and business, *The Chem. Eng.* (Sept.), 404
Zuiderweg, F. J. (1982). Sieve trays – a view on the state of the art, *Chem. Engng. Sci.* **37**, (10), 1441
Zuiderweg, F. J. (1983). Marangoni effect in distillation of alcohol–water mixtures, *Chem. Eng. Res. Des.* **61**, (Nov.), 388
Zuiderweg, F. J. (1984). Personal communication
Zuiderweg, F. J. and Harmens, A. (1958). The influence of surface phenomena on the performance of distillation columns, *Chem. Engng. Sci.* **9**, 89
Zuiderweg, F. J., Hofhuis, P. A. M. and Kuzniar, J. (1984). Flow regimes on sieve trays: the significance of the emulsion flow regime, *Chem. Eng. Res. Des.* **62,** Jan., 39
Zuiderweg, F. J., de Groot, J. H., Meeboer, B. and van der Meer, D. (1969). Scaling up distillation plates, *Inst. Chem. Engrs. Symp. Series No. 32*, p. 5:78

Index